5

AF768155

NANOTECHNOLOGY SCIENCE AND TECHNOLOGY

NANOTECHNOLOGY

SELECT ASSESSMENTS OF THE NATIONAL NANOTECHNOLOGY INITIATIVE

NANOTECHNOLOGY SCIENCE AND TECHNOLOGY

Additional books in this series can be found on Nova's website under the Series tab.

Additional E-books in this series can be found on Nova's website under the E-book tab.

NANOTECHNOLOGY SCIENCE AND TECHNOLOGY

NANOTECHNOLOGY

SELECT ASSESSMENTS OF THE NATIONAL NANOTECHNOLOGY INITIATIVE

STEVEN M. GOLDFARB
AND
HUGH MORGAN
EDITORS

nova publishers
New York

Copyright © 2013 by Nova Science Publishers, Inc.

All rights reserved. No part of this book may be reproduced, stored in a retrieval system or transmitted in any form or by any means: electronic, electrostatic, magnetic, tape, mechanical photocopying, recording or otherwise without the written permission of the Publisher.

For permission to use material from this book please contact us:
Telephone 631-231-7269; Fax 631-231-8175
Web Site: http://www.novapublishers.com

NOTICE TO THE READER

The Publisher has taken reasonable care in the preparation of this book, but makes no expressed or implied warranty of any kind and assumes no responsibility for any errors or omissions. No liability is assumed for incidental or consequential damages in connection with or arising out of information contained in this book. The Publisher shall not be liable for any special, consequential, or exemplary damages resulting, in whole or in part, from the readers' use of, or reliance upon, this material. Any parts of this book based on government reports are so indicated and copyright is claimed for those parts to the extent applicable to compilations of such works.

Independent verification should be sought for any data, advice or recommendations contained in this book. In addition, no responsibility is assumed by the publisher for any injury and/or damage to persons or property arising from any methods, products, instructions, ideas or otherwise contained in this publication.

This publication is designed to provide accurate and authoritative information with regard to the subject matter covered herein. It is sold with the clear understanding that the Publisher is not engaged in rendering legal or any other professional services. If legal or any other expert assistance is required, the services of a competent person should be sought. FROM A DECLARATION OF PARTICIPANTS JOINTLY ADOPTED BY A COMMITTEE OF THE AMERICAN BAR ASSOCIATION AND A COMMITTEE OF PUBLISHERS.

Additional color graphics may be available in the e-book version of this book.

Library of Congress Cataloging-in-Publication Data

ISBN: 978-1-62257-596-1

Published by Nova Science Publishers, Inc. ✛ New York

CONTENTS

PREFACE

This book provides an overview of the National Nanotechnology Initiative (NNI) which was launched in 2001, with an initial budget of $500 million. The NNI is a crosscutting Federal program designed to coordinate U.S. investment in research and development (R&D) activities in nanoscale science, engineering, technology, and related efforts. The NNI has four broad objectives: (1) to advance world-class nanotechnology research and development; (2)to foster the transfer of new technologies into products for commercial and public benefit; (3)to develop and sustain educational resources, a skilled workforce, and the supporting infrastructure and tools to advance nanotechnology; and (4)to support the responsible development of nanotechnology. The NNI has had a "catalytic and substantial impact" on the growth of the U.S. nanotechnology industry and as a result, the United States is today, the global leader in this exciting and economically promising field of research and technological development.

Chapter 1 – Report on Nanotechnology: Improved Performance Information Needed for Environmental, Health, And Safety Research, of United States Government Accountability Office, on May 2012.

Chapter 2 – Report to the President And Congress on the Fourth Assessment of the National Nanotechnology Initiative of Executive Office of the President, dated April 2012.

In: Nanotechnology
Editors: S.M. Goldfarb, H. Morgan
ISBN: 978-1-62257-596-1
© 2013 Nova Science Publishers, Inc.

Chapter 1

REPORT TO THE PRESIDENT AND CONGRESS ON THE FOURTH ASSESSMENT OF THE NATIONAL NANOTECHNOLOGY INITIATIVE*

President's Council of Advisors on Science and Technology

WASHINGTON, D.C. 20502
President Barack Obama
The White House
Washington, DC 20502

Dear Mr. President,

We are pleased to send you this "Report to the President and Congress on the Fourth Assessment of the National Nanotechnology Initiative," prepared by the President's Council of Advisors on Science and Technology (PCAST). This report fulfills PCAST's responsibilities under the 21st Century Nanotechnology Research and Development Act (Public Law 108-153) and Executive Order 13349 to provide periodic updates to Congress.

The report finds that the NNI—which has provided $16 billion in investments by 26 Federal agencies over the life of the initiative—has had a

---

* This is an edited, reformatted and augmented version of Executive Office of the President, dated April 2012.

"catalytic and substantial impact" on the growth of the U.S. nanotechnology industry and should be continued. Further, the report finds that in large part as a result of the NNI the United States is today, by a wide range of measures, the global leader in this exciting and economically promising field of research and technological development. PCAST applauds the increased efforts of the National Nanotechnology Coordinating Office in the area of commercialization and coordination with industry and in the release of a focused research strategy for addressing environmental, health, and safety (EHS) implications of nanotechnology. In addition, the NNI maintains a strong portfolio of research on the societal implications of nanotechnology.

However, the report also notes that additional efforts are needed in four areas: strategic planning, program management, metrics for assessing impact, and increasing support for research on environmental, health, and safety issues associated with nanotechnology. Continued lack of attention to these concerns will make it harder for the U.S. to maintain its leadership role in the commercialization of nanotechnology. The full PCAST discussed and approved this report at its most recent public meeting on March 9, 2012. We appreciate your interest in this important field of work and sincerely hope that you find this report useful.

John P. Holdren
Co-Chair

Eric Lander
Co-Chair

EXECUTIVE SUMMARY

The National Nanotechnology Initiative (NNI) is a crosscutting Federal program designed to coordinate U.S. investment in research and development (R&D) activities in nanoscale science, engineering, technology, and related efforts across 26 agencies and programs. This is the fourth review of the NNI by the President's Council of Advisors on Science and Technology (PCAST) since the council was designated in 2004 as the National Nanotechnology Advisory Panel tasked with reviewing the initiative. The Federal Government has proposed $1.8 billion of funding in fiscal year (FY) 2013 for 15 agencies with budgets dedicated to nanotechnology research and development. The FY 2013 request represents total funding of $18 billion over the life of the

Initiative. Nearly 75 percent of this funding goes to three Program Component Areas: Fundamental Nanoscale Phenomena and Processes, Nanomaterials, and Nanoscale Devices and Systems. The NNI continues to support a strong and growing portfolio of research on the societal implications of nanotechnology, nanotechnology education, and public outreach. The President's 2013 budget includes a total of $306 million—a 24-percent increase compared to 2011 actual spending—for three Nanotechnology Signature Initiatives: Nanotechnology for Solar Energy Collection and Conversion; Sustainable Nanomanufacturing: Creating the Industries of the Future; and Nanoelectronics for 2020 and Beyond. These initiatives foster meaningful interagency collaboration and serve as springboards for the rapid advancement of nanoscience and technology toward commercialization.

Progress on 2010 Recommendations

PCAST's 2010 review of the NNI included recommendations in the categories of program management; nanotechnology outcomes; and environment, health, and safety (EHS). In this review, PCAST found that Federal agencies and offices involved in the NNI have made substantial progress in many areas, but little progress on some of the key recommendations PCAST made in 2010. On the positive side, the NNI has made progress in these areas:

- The National Nanotechnology Coordination Office (NNCO) expanded efforts in the area of commercialization and released a focused research strategy for addressing the EHS implications of nanotechnology.
- The NNCO has developed the Industry and State Liaison position to serve as a point of contact for the private sector.
- The National Science and Technology Council (NSTC) Nanoscale Science, Engineering, and Technology (NSET) Subcommittee's Nanomanufacturing, Industry Liaison, and Innovation Working Group is developing mechanisms to incorporate industrial input in NNI planning through public-private partnerships and is developing an agenda that focuses on job creation and state outreach.
- The Department of Energy initiated programs that include industrial partners to overcome technological barriers to nanotechnology commercialization.

- The National Institute of Standards and Technology plans to start the Advanced Manufacturing Technology Consortia in FY 2013 to speed up the development and commercialization of new products and services, including nanotechnology.
- The National Institutes of Health created the National Center for Advancing Translational Sciences to accelerate translation of promising technologies and clinical studies.
- The NSET Subcommittee created a standalone EHS strategy that reflects evolving research needs and the strategic research plans of three relevant agencies.

While these advances are encouraging, PCAST is concerned that little progress has been made since 2010 in four broad areas:

- *Strategic Planning.* While the NSET Subcommittee in 2011 produced a "National Nanotechnology Initiative Strategic Plan," individual agency contributions lack the cohesion of an overarching framework, and there is no clear connection between the goals and objectives of the NNI strategic plan with those of individual agencies.
- *Program Management.* PCAST is concerned that the agency representatives appointed to the NSET Subcommittee do not have a level of authority within their agencies to influence budget allocations needed to meet NNI objectives. In addition, mechanisms to solicit and act upon advice from outside of the Federal Government are still inadequate, as is the level of funding and capacity of the NNCO leadership to support the agencies in implementing programs that align with the NNI strategic plan.
- *Metrics.* The lack of clear metrics for assessing the impacts of Federal investments in nanotechnology remains a concern. Little appears to have been done to spur the development of metrics needed to determine the economic outcomes of the initiative.
- *Environmental Health and Safety.* PCAST is concerned that there is still a lack of integration between nanotechnology-related EHS research funded through the NNI and the kind of information policymakers need to effectively manage potential risks from nanomaterials.

In August 2011, the first cohort of National Science Foundation (NSF) Nanoscale Science and Engineering Centers was sunsetted. These centers have served as highly visible evidence of robust and coordinated U.S. support for nanotechnology in critical subject areas. Each of the old centers has been encouraged to seek additional funding to sustain its efforts beyond the lifetime of its NSF grant. Whether new or old Centers are funded, all future Centers should continue to exemplify an equal level of leadership, infrastructure, programs for training, and educational and public outreach and play a key role in the overarching strategic plan for nanotechnology.

Recommendations

PCAST's recommendations in the four broad areas still of concern are summarized here:

Strategic Planning

- The NNCO in partnership with the Office of Science and Technology Policy (OSTP) should work with the agencies to develop agency implementation plans for achieving the goals and objectives outlined in the 2011 NNI strategic plan.
- Participating agencies should ensure that senior agency officials capable of influencing funding decisions are participating fully and personally in strategic planning activities of the NSET. Officials at this level, in contrast with representatives active at the program or office level, could more effectively drive agency planning and budget allocations to meet NNI strategic directions.
- The Nanotechnology Signature Initiatives should be fully supported in NNI budgets. To this end, PCAST recommends that the Office of Management and Budget increase funding to these Initiatives.
- The NSET Subcommittee should create Nanotechnology Signature Initiatives in other priority areas such as homeland security, national defense, and human health.

Program Management

OSTP should facilitate the following:

- Appoint the NNCO director as co-chair of the NSET Subcommittee of the NSTC.
- Change the requirement that the NNCO director must come from within the Federal Government to allow external, non-Federal experts the opportunity to direct the NNCO.
- Create a standing PCAST Nanotechnology Steering Committee of experts from industry, academia, and civil society to provide more frequent and in-depth guidance to the overall initiative and to the signature initiatives.
- Dedicate 0.3 percent of NNI funding to the NNCO to ensure the appropriate staffing and budget to effectively develop, monitor, and assess NNI programs.
- Work with the NNCO director to develop a plan for increasing the NNCO budget in line with its new responsibilities.

Metrics

- Agencies should develop a mission-appropriate definition of nanotechnology that enables tracking specific nanotechnology investments supported at the program level. The definition and funding details should be published in agency implementation plans to promote clarity.
- The NNCO should track the development of metrics for quantifying the Federal nanotechnology portfolio and implement them to assess NNI outputs

Environmental, Health, and Safety (EHS)

The NSET should:

- Establish high-level, cross-agency authoritative and accountable governance of Federal nanotechnologyrelated EHS research so that the knowledge created as a result of Federal investments can better inform policy makers.
- Increase investment in cross-cutting areas of EHS that promote knowledge transfer such as informatics, partnerships, and instrumentation development.

INTRODUCTION

In 2001 the Federal Government launched the National Nanotechnology Initiative (NNI) with an initial budget of $500 million. The NNI is a crosscutting Federal program designed to coordinate U.S. investment in research and development (R&D) activities in nanoscale science, engineering, technology, and related efforts across 26 agencies and programs, 15 of which have budgets dedicated to nanotechnology research and development. The NNI has four broad objectives: (1) to advance world-class nanotechnology research and development; (2) to foster the transfer of new technologies into products for commercial and public benefit; (3) to develop and sustain educational resources, a skilled workforce, and the supporting infrastructure and tools to advance nanotechnology; and (4) to support the responsible development of nanotechnology.[1] Many of these goals may take decades to develop. As President Clinton explained in his January 21, 2000, speech at the California Institute of Technology, "Some of our research goals will take twenty or more years to achieve. But that is why, precisely why...there is such a critical role for the Federal Government."[2]

The NNI is managed within the framework of the Cabinet-level National Science and Technology Council (NSTC) through the Nanoscale Science, Engineering, and Technology (NSET) Subcommittee. The NSET is composed of representatives from participating agencies and coordinates the planning, budgeting, program implementation, and review of the initiative and has four working groups dedicated to key NNI activities. The National Nanotechnology Coordination Office (NNCO) provides technical and administrative support to the NSET Subcommittee, including the preparation of multiagency planning, budget, and assessment documents. The NNCO also serves as the primary point of contact for agencies participating in the NNI as well as for public engagement and outreach. (An NNI organizational chart is found in Appendix B.)

The 21st Century Nanotechnology Research and Development Act of 2003 (Public Law 108-153) calls for a National Nanotechnology Advisory Panel (NNAP) to periodically review the NNI. The President's Council of Advisors on Science and Technology (PCAST) was designated as the NNAP in 2001 and subsequently re-instated in 2004 by Executive Order and reviewed the NNI in 2005, 2008, and 2010. This report represents PCAST's fourth review of the NNI as the NNAP.

What Is Nanotechnology?

Nanotechnology is the control and restructuring of matter at the nanoscale, in the size range of approximately 1–100 nanometers, in order to create materials, devices, and systems with fundamentally new properties and functions due to their small structure. Encompassing nanoscale science, engineering, and technology, nanotechnology involves imaging, measuring, modeling, and manipulating matter at this length scale. A nanometer is one-billionth of a meter. One sheet of paper is about 100,000 nanometers thick; a single gold atom is about a third of a nanometer in diameter.

Table 1. National nanotechnology initiative funding for select agencies 2009–2013 (dollars in millions)

Agency	2009	2009 Recovery	2010	2011	2012 Estimate**	2013 Proposed
Department of Energy (DOE)	333	293	374	346	315	443
National Science Foundation (NSF)	409	101	429	485	426	435
Department of Health and Human Services (HHS)/National Institutes of Health (NIH)	343	73	457	409	410	409
Department of Defense (DOD)	459	—	440*	425	361	289
Department of Commerce (DOC)/ National Institute of Standards and Technology (NIST)	93	43	115	96	95	102
National Aeronautics and Space Administration (NASA)	14	—	20	17	23	22
Environmental Protection Agency (EPA)	12	—	18	17	18	19

* Includes $75 million in congressionally directed funding that is outside the NNI plans.

** Based on FY 2012 appropriated levels.

* In the initial public release of this report on April 27, 2012 this sentence contained several typos which were corrected and the updated version of the report was released on May 4, 2012.

Since 2001, the Federal Government has committed over $16 billion to the NNI, and has proposed $1.8 billion of funding in fiscal year (FY) 2013 for the 15 participating agencies with budgets dedicated to nanotechnology research and development. This request represents a $146 million decrease from 2010 funding levels.* At least in part, this investment has been made to ensure that the United States continues to maintain a global leadership position in nanotechnology research, development, and commercialization. There are three ways to consider how NNI funds are distributed: by agency, by Program Component Area (PCA), or by Nanotechnology Signature Initiative (NSI). Table 1 provides recent NNI funding for the seven agencies with the highest funding. Four of the seven agencies (DOE, NSF, NASA and EPA) show an increase in funding between the 2010 actual and 2013 proposed budgets.*

NNI investments are distributed across eight PCAs, as determined collaboratively among the White House Office of Science and Technology Policy (OSTP), the Office of Management and Budget (OMB) and the participating agencies. The PCAs are thematic areas that represent the organizational framework under which NNI research and major activities are grouped. The eight PCAs are:

- PCA 1: Fundamental Nanoscale Phenomena and Processes
- PCA 2: Nanomaterials
- PCA 3: Nanoscale Devices and Systems
- PCA 4: Instrumentation Research, Metrology, and Standards for Nanotechnology
- PCA 5: Nanomanufacturing
- PCA 6: Major Research Facilities and Instrumentation Acquisition
- PCA 7: Environment, Health, and Safety
- PCA 8: Education and Societal Dimensions

In addition to the PCAs, the NNI has established three NSIs to foster meaningful interagency collaboration:

- Nanotechnology for Solar Energy Collection and Conversion
- Sustainable Nanomanufacturing: Creating the Industries of the Future
- Nanoelectronics for 2020 and Beyond

Unlike the agency budgets or PCAs, NSIs do not provide a complete crosscut of all NNI activity; there is much support for nanotechnology that is not captured within an NSI. They represent areas within the overall NNI activity where there is particular opportunity for rapid progress through collaborative actions across agencies. These initiatives are consistent with the President's Strategy for American Innovation[3] and are intended to serve as springboard programs for the rapid advancement of nanoscience and technology toward commercialization.[4] The President's 2013 Budget includes a total of $306 million for the initiatives, with $112 million for

Nanotechnology for Solar Energy Collection and Conversion, $84 million for Sustainable Nanomanufacturing, and $110 million for Nanoelectronics for 2020 and Beyond.

Several trends in budget allocations across PCAs (Figure 1), which are highlighted in the NNI FY 2013 Supplement to the President's Budget, represent significant changes since the publication of PCAST's previous review of the NNI.[5]

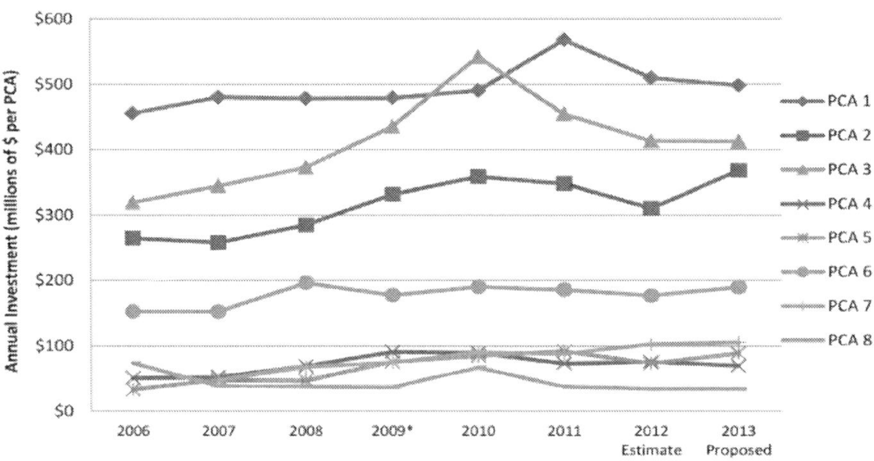

Source: NNI Supplement to the President's Budget (2007-2013).
* Not including ARRA funding.

Figure 1. Annual NNI investment per PCA, 2006-2013.

Funding under PCAs 1, 2 and 3, which account for nearly 75 percent of the total NNI funding request, has decreased 8.0 percent from $1.39 billion in 2010 to $1.28 billion in the 2013 request. Agency officials report that the NNI investment in nanomanufacturing (PCA 5) rose 4.8 percent from $84.8 million in 2010 to $88.9 million in 2013. This captures increased investments associated with the NSI on Sustainable Nanomanufacturing as well as increased investments at the NSF ($29.3 million in the 2010 actual budget to $52.8 in the 2013 request). While fundamental research (PCA 1) remains the largest single NNI investment category ($498 million in the 2013 budget), the more applied research in nanodevices and systems (PCA 3) and in nanomanufacturing (PCA 5) now totals over $500 million combined, as some areas of nanotechnology mature and applications develop. The funding level for PCA 7, Environment, Health, and Safety (EHS), increased by 16 percent, from $90.2 million in 2010 to $105 million in the 2013 request. As initial

Federal investments in nanotechnology mature, more attention is being paid to the outputs from these research activities. Early research in nanoscience and nanotechnology has resulted in both publication and patenting, which has further enabled researchers to investigate the effects of these national investments on an international scale. Publication rates in nanotechnology, which can be used as an indicator of research productivity, show that the United States is falling in its share of total nano-related publications but still leads in publishing in three top science journals (*Science*, *Nature*, and *Proceedings of the National Academy of Sciences*) (Figure 2 and Appendix D).

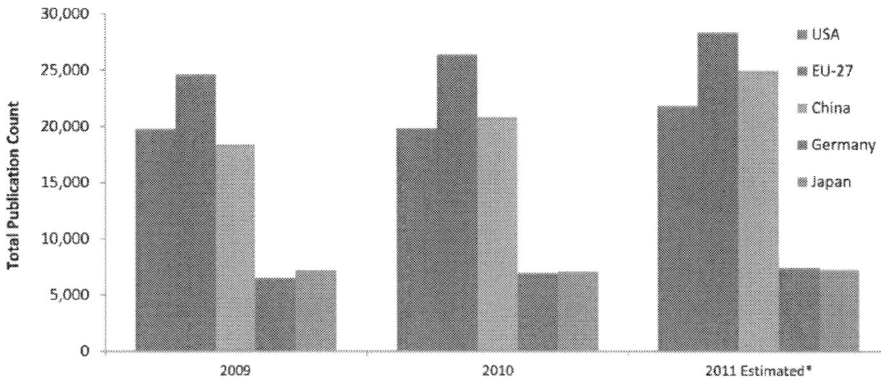

Source: L. Kay, P. Shapira, J. Youtie, and A. L. Porter, Program on Nanotechnology Research and Innovation System Assessment Georgia Institute of Technology, Atlanta, Georgia, with support from the National Science Foundation (NSF) through the Center for Nanotechnology in Society (Arizona State University; Award No. 0937591) [Data as January 2012].

Note: Search based on S. Arora, A. L. Porter, Y. Youtie, and S. Shapira, Capturing new developmnets in nanotechnology scientific output: A search strategy for publication records. STIP White Paper, Atlanta, Georgia, 2011.

* Annualized publication counts based on number of country-level publications identified as of 22 August 2011. Additional 5% added to account for publications expected to be released in the next calendar year (i. e., in early 2012).

Figure 2. Publication counts for nanotechnology articles in Web of Science, 2009–2011 (SCI, SSCI, and A&HCI).

Fgure 2 provides publication counts for nanotechnology articles from 2009 to 2011.[6] Publication counts are reported for the top five nanotechnology publication-producing countries for all journals collected in Thomson Reuters Web of Science (Science Citation Index [SCI], Social Science Citation Index [SSCI], Arts and Humanities Citation Index [A&HCI]).

The United States, the 27 European Union (EU-27) member nations, and China outpace the rest of the world in numbers of publications produced. Of additional importance, the number of papers produced globally that cite nano-publications with U.S. authorship remains high. Through 2010, the United States held the largest percentage of nanotechnology citations. Citation counts corresponding to nanotechnology publication counts presented in Figure 2 appear in Appendix Figures D-1 and D-2. Researchers and firms from the United States continue to claim the highest number of nanotechnology priority patents internationally, outnumbering the second place EU-27 bloc by almost 3,000 filings in 2005–2009 (Figure 4).

However, South Korean and Chinese nanotechnology patent applications (measured by assignee country) are increasing quickly. Figures 3 and 4 show counts of patent applications by assignee country over five-year increments. Figure 3 includes all patent applications filed, whereas Figure 4 restricts assignee patent counts to only documents that report priority filing. A priority patent is one in which the intellectual property holder is recognized as the first to file and therefore receives exclusive, if time-limited, rights to the intellectual property internationally.

A more accurate illustration of global nanotechnology intellectual property rights requires considering that priority patent applications indicate which countries will successfully hold the intellectual property pertaining to a technology patent family. This is particularly relevant when analyzing the discrepancies between counts of all patent applications and priority patent applications. Priority patent figures show a sharp fall in China's ranking from first place when considering all patent applications to eighth place when considering priority patents only. Moreover, patent counts are based on the analysis of patent documents that refer to nanotechnology-relevant topics (such as methods, materials, or processes) and do not yet indicate whether or not the technologies are ultimately commercially viable.

The United States continues to lead in the race to nanotechnology commercialization; it holds the highest number of priority patents by almost 5,000 filings in the time period 2005–2009. Figure 4 also aggregates European countries into one category (EU 27), and it illustrates counts of priority patent applications in nanotechnology globally, considering the importance of the EU-27 as a single community. Additional nanotechnology bibliometrics data appear in Appendix D.

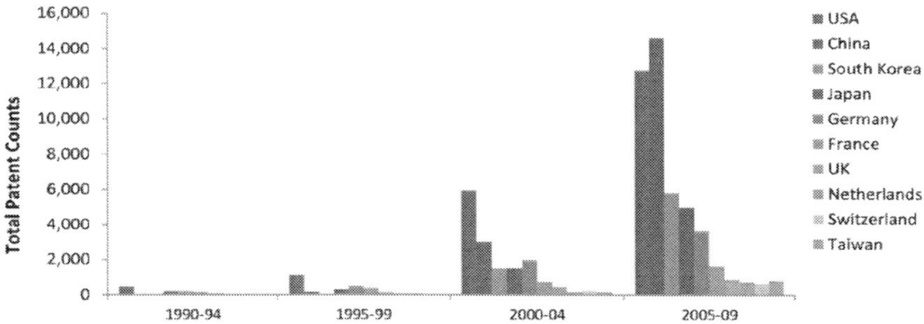

Source: L. Kay, P. Shapira, J. Youtie, and A. L. Porter, Program on Nanotechnology Research and Innovation System Assessment Georgia Institute of Technology, Atlanta, Georgia, with support from the National Science Foundation (NSF) through the Center for Nanotechnology in Society (Arizona State University; Award No. 0937591) [Data as of January 2010]. PATSTAT stand for the Patent Statistical Database, which is maintained by the European Patent Office.

Note: Based on Definition of "nanotechnology" by A. L. Porter, J. Youtie, P. Shapira, and D. Schoeneck, "Refining Search Terms for Nanotechnology," *Journal of Nanoparticle Research,* 10 (5): 715-728.

Figure 3. Counts of patent applications in nanotechnology in PATSTAT, by year and assignee country, 1990–2009.

Source: L. Kay, P. Shapira, J. Youtie, and A. L. Porter, Program on Nanotechnology Research and Innovation System Assessment Georgia Institute of Technology, Atlanta, Georgia, with support from the National Science Foundation (NSF) through the Center for Nanotechnology in Society (Arizona State University; Award No. 0937591) [Data as of January 2010]. PATSTAT stands for the Patent Statistical Database, which is maintained by the European Patent Ofiice.

Note: Based on definition of "nanotechnology" by A. L. Porter, J. Youtie, P. Shapira, and D. Schoeneck, "Refining Search terms for Nanotechnology," *Journal of Nanoparticle Research,* 10 (5): 715-728. Proirity applications are patent documents in PATSTAT that report priority sequence of "1".

Figure 4. Counts of priority patent applications in nanotechnology in PATSTAT, by year and assignee country, 1990–2009.

INITIATIVE DEVELOPMENTS SINCE 2010

PCAST's 2010 assessment found that "U.S. leadership in nanotechnology is threatened by several aggressively investing competitors such as China, South Korea, and the European Union. In response to this threat, the report recommends a number of changes in Federal programs and policies, with the goal of assuring continued U.S. dominance in the decade ahead."[7] The report presented recommendations in three broad categories—program management, nanotechnology outcomes, and EHS—each with a number of specific recommendations for NNCO and Federal agencies. This assessment addresses progress toward these specific recommendations through an examination of government documents, expert presentations, and evidence provided by NNCO and agency officials.

Graphene: The Material of the Future

The 1985 discovery at Rice University of a new form of pure carbon, the soccer-ball shaped C60 molecule known as buck-minsterfullerene, renewed interest in carbon research and garnered the Nobel Prize in Chemistry just eleven years later.

The spheroidal molecule's tubular relatives, carbon nanotubes have generated even more attention and focus due to their extraordinary physical and chemical properties. Now, there's excitement over the two-dimensional hexagonal, crystalline carbon material called graphene. While scientists have known about graphene for decades, it was only in 2004 that grapheme flakes were isolated and imaged and their electrical conductivity characterized. For this achievement, Andre Geim and Konstantin Novoselov were awarded the 2010 Nobel Prize in Physics.

At one atom thick, graphene is ultimately thin, nearly completely transparent, and super strong yet flexible. Graphene is more than one hundred times stronger than steel of an equivalent thickness, ten times more thermally conductive than copper, and slightly more electrically conductive than copper. Its electrical conductivity recommends it as a potential replacement for silicon in transistors; its smaller size profile would allow a higher transistor density on a chip, enabling even greater miniaturization of computer components than is possible with silicon-based technologies.

The combination of transparency, thinness, and electrical conductivity has led to predictions that graphene will ultimately replace the expensive and fragile indium-tin-oxide material used in many touch screens. Add in graphene's flexibility and one can envision it enabling the development of computer displays that can be rolled up like a newspaper.

For these possibilities to be realized, however, several material challenges must be addressed. Specifically needed is a method for scaling the production of graphene and a way to alter the graphene so that it acts as a semiconductor for use in computers. The Federal Government has sponsored 20 Small Business Innovation Research (SBIR/STTR) awards relating to graphene research in the last three years. Of these awards, two sponsor Phase II research into the development of specific material processing and characterization technologies for scaling graphene production. Phase I awards focus on overcoming barriers to commercial application of graphene in technologies including supercapacitors for energy storage in hybrid electric vehicles and sensors for chemical explosives.

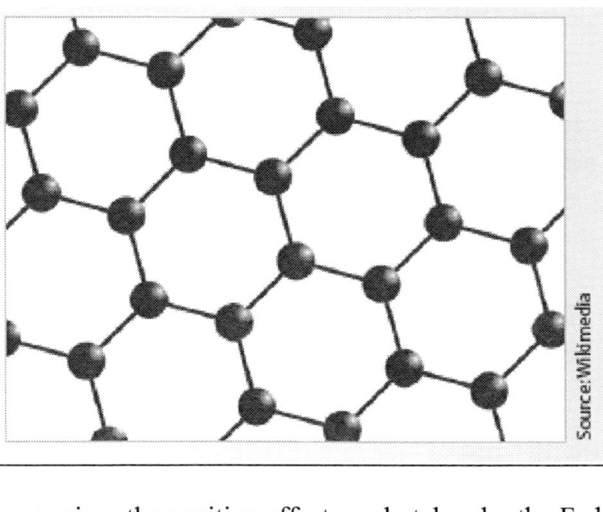

PCAST recognizes the positive efforts undertaken by the Federal agencies and offices involved in the NNI toward addressing the recommendations made in the 2010 PCAST assessment (see Appendix E for the full list of recommendations). Of particular note are the expanded efforts of the NNCO in the area of commercialization and the release of a focused research strategy for addressing the EHS implications of nanotechnology. Additionally, there have been many advances coming out of the federally-funded programs some of which are already moving on to commercialization phases. In this section, we highlight three such advances beginning with the Nobel prize winning graphene example.

In support of PCAST's 2010 recommendations, DOE has initiated programs that include industrial partners to overcome technological barriers to nanotechnology commercialization. NIST plans to start the Advanced Manufacturing Technology Consortia in FY 2013 to speed up the development and commercialization of new products and services, including nanotechnology, and NIH has created the National Center for Advancing Translational Sciences to accelerate translation of promising technologies and clinical studies. Finally, the NNCO has developed the Industry and State Liaison (ISL) position to serve as a point of contact for the private sector. The NSET Nanomanufacturing, Industry Liaison, and Innovation (NILI) Working Group is developing mechanisms to incorporate industrial input in NNI planning through public-private partnerships and is developing an agenda that focuses on job creation and state outreach.

Liquidia: Nano-enabled Vaccines and Therapeutics

Liquidia Technologies in Research Triangle Park, North Carolina, is developing highly precise nanoparticle-based vaccines and therapeutics for the prevention and treatment of human disease. It leverages techniques from the semiconductor industry to fabricate particles in a highly regular manner that can mimic the properties of viruses and other biological entities. With its proprietary PRINT technology, Liquidia has exquisite control over particle size, shape, and composition and can prepare a wide variety of highly functional materials for drug delivery and adjuvant therapy purposes. Its first products in vaccines (influenza and malaria) and its emerging products for treating respiratory disease (COPD and pulmonary hypertension) are enabled by the recent transition of its PRINT nanoparticle technology to a manufacturing process that is compliant with current Good Manufacturing Practices (cGMP). Liquidia was founded in 2004 by Professor Joseph M. DeSimone from the University of North Carolina at Chapel Hill, whose research has been sponsored in part by the National Cancer Institute's Alliance for Nanotechnology in Cancer Centers of Cancer Nanotechnology Excellence (CCNE), NSF EArly-concept Grants for Exploratory Research (EAGER), and the NIH Director's Pioneer Award in 2009.

Liquidia is on a rapid path to the clinic and has secured the first equity investment by the Bill & Melinda Gates Foundation and resources from a variety of major pharmaceutical companies. The company is backed by venture capital investment, led by Canaan Partners and New Enterprise Associates, both located in Menlo Park, California.

PCAST specifically notes the laudable effort undertaken by the NSET Subcommittee to create a standalone EHS strategy that is reflective of evolving research needs and strongly coupled to the strategic research plans of three relevant agencies. *NNI Environmental, Health, and Safety Research Strategy* was released in October 2011 by the Nanotechnology Environmental and Health Implications (NEHI) working group of the NSET Subcommittee. The strategy made significant progress toward the 2010 PCAST recommendation to provide cross-agency guidance by outlining plans for identifying research areas that align with agency missions. The plans utilize the NNI EHS Coordinator as a central facilitator across the NSET agencies and internationally and establish more frequent assessments of progress made toward the strategic goals by the NEHI working group.[8] The EHS Research Strategy aligns well with the findings of an independent panel of experts convened by National Research Council, whose January 2012 report, *A Research Strategy for Environmental, Health, and Safety Aspects of Engineered Nanomaterials (ENMs),*[9] was already under review when the EHS Research Strategy was released. Specific areas of alignment between the two documents include recognition of the importance of a life cycle approach to assessing risks, the need for more research on human and environmental exposure to nanomaterials, better tools for measuring and tracking nanomaterials, and the need for cross-cutting informatics infrastructure for nanotechnology-related EHS research.

Additionally, the NNI continues to support a strong and growing portfolio of research on the societal implications of nanotechnology, nanotechnology education, and public outreach. In late 2010, the NSF released *Nanotechnology Research Directions for Societal Needs in 2020*, including a chapter on the responsible governance of nanotechnology for societal development. The NSF continues to support two Centers for Nanotechnology in Society dedicated to research on the societal dimensions of nanotechnology and public outreach. Currently, NSF supports the training and education of roughly 10,000 students and teachers in nanoscale science and engineering. NSF also funds the development of new curricula for nanotechnology education and is expanding the outreach of the National Center for Nanotechnology Applications and Career Knowledge.

Several major developments since the 2010 PCAST report will likely be important in the coming years. Two of these developments are examined in the following sections.

Sunset of the First Class of Nanoscale Science and Engineering Centers

One notable development since the publication of the 2010 PCAST report is the sunsetting of the first class of National Science Foundation Nanoscale Science and Engineering Centers (NSECs) in August 2011.[10] These centers were created in 2001 with an initial five-year award and were renewed in 2006 for an additional five years. In May 2011, a team of NSF-funded researchers published a report entitled, "Assessment of Fifteen Nanotechnology Science and Engineering Centers' Outcomes and Impacts: Their contribution to NNI Objectives and Goals".[11] This team found evidence of NSEC contributions to world-class nanotechnology R&D through citation of the NSECs' published outputs by global research organizations, the presence of NSEC publications in the highest impact journals, citations of early NSEC papers being among the most highly cited papers in the field, and a very significant spillover effect of center research into adjacent fields that underscores the interdisciplinary nature of NSEC research. The team also concluded that the centers had been successful in fostering the transfer of new technologies into products, citing the rich and collaborative links between NSECs and industrial partners, and the fact that a core of top high-technology firms came to rely on the centers as an R&D resource. While this research studied 15 NSECs, the report singled out several within the initial cohort of six centers for special recognition in key areas such as research performance, commercialization of NSEC research, education, and responsible development.

Workforce Development

With the support of the NSF's Advanced Technology Education (ATE) program, Penn State has developed a nation-wide partnership of research universities and community colleges that is bringing meaningful core-skills nanotechnology workforce education to technical and community colleges across the United States. This partnership, the NSF National Nanotechnology Applications and Career Knowledge (NACK) Network, fosters (1) resource sharing among community colleges and research universities for nanotechnology workforce development, (2) the availability of course materials, for web or in-class use, covering a core-set of industry-recommended nanotechnology skills and (3) broad student preparation for careers in the wide spectrum of industries utilizing micro- or nanotechnology. NACK has created and offers continually updated, free-of-charge core-skills course lecture and lab materials, web-accessible equipment capability, and faculty development workshop curricula. Since the inception of the nationwide effort in 2008, NACK research university-community college partnership hubs have been set-up and are functioning in Puerto Rico, New York, Indiana, Minnesota, Texas, and Washington State. Others are underway and these are in addition to the hub comprised of 30 Pennsylvania schools and funded by the State of Pennsylvania since 1998. To-date, there have been over 800 graduates from the nanotechnology core-skill classes offered by the NACK hubs, 20,881 web downloads of NACK educational materials, and 957 educators who have completed professional development workshops. The Penn State nanotechnology workforce development programs began as a Pennsylvania-focused activity with the founding of Pennsylvania Nanofabrication Manufacturing Technology (NMT) Partnership funded by the State in 1998. In 2003 the additional component of an NSF ATE regional center for nanotechnology workforce education was added. In 2008 this NSF ATE activity evolved into the NACK Network nationwide workforce development partnership.

By creating education pathways from high school to skilled manufacturing careers across the country, the NACK Network is working to train the U.S. nanotechnology manufacturing workforce.

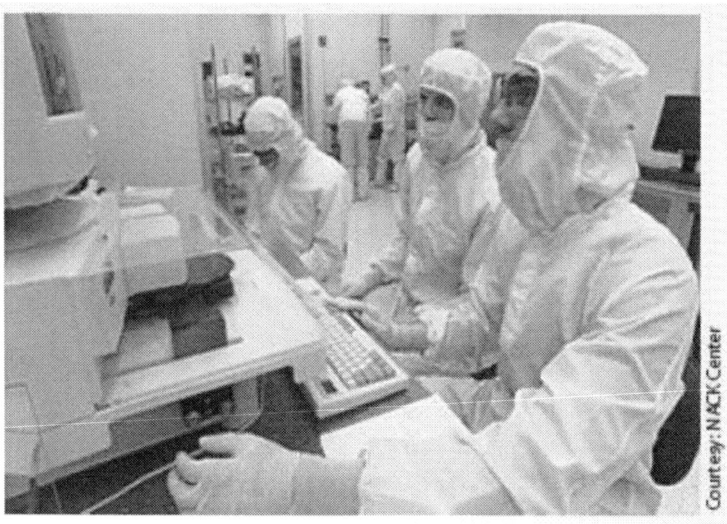

The centers serve as highly visible evidence within the international community of robust and coordinated U.S. support for nanotechnology in critical subject areas. Each of the old centers has been encouraged to competitively seek funding to sustain its efforts beyond the lifetime of its NSF grant. However, it is also important that any new cohort of centers exemplify an equal level of leadership, support of infrastructure, programs for training, and educational and public outreach as well as play a key role in the overarching strategic plan for nanotechnology as the old centers have.

International Developments

The Federal Government continues to invest more in nanotechnology R&D than any other single country (Figure 5). To date more than $16 billion has been invested in U.S. nanoscience and nanotechnology via the NNI. There has been concern that in addition to China, South Korea, and other early movers, the Russian Nanotech Corporation (RUSNANO) is now also rising as a major player, second only to the United States in its nanotechnology R&D

spending. According to Lux Research, RUSNANO increased its funding by nearly 40 percent to $1.05 billion and has plans to increase even further to nearly $1.5 billion by 2015.

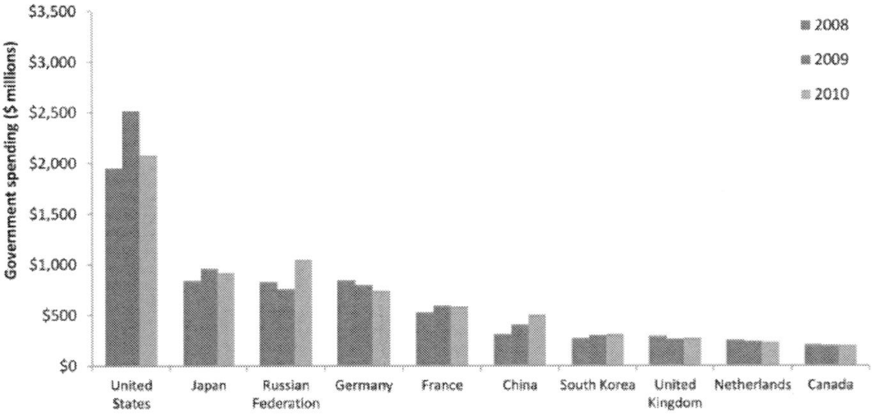

Source: A. Xue and D. Hwang, "Nanotechnology Funding: Corporations Gran the Reins," Lux Research, April 2011.

Figure 5. Global governmental nanotechnology spending for top ten countries, 2008–2010.

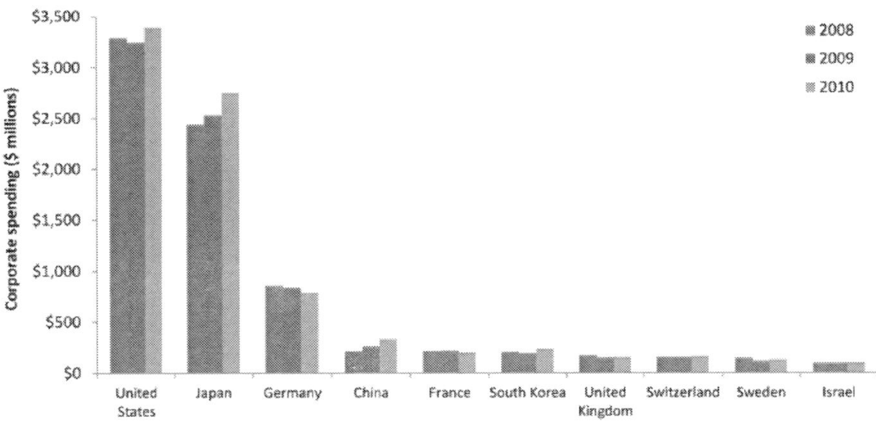

Source: A. Xue and D. Hwang, "Nanotechnology Funding: Corporations Gran the Reins," Lux Research, April 2011.

Figure 6. Corporate nanotechnology spending for top ten countries, 2008–2010.

Despite a significant surge in Russian spending in 2010 (Figure 5), bibliometrics data do not indicate a commensurate increase in scholarly or commercial output. It may be premature to assess the potential impact of Russian investment in nanoscale research and development. Within Europe, Germany is currently outspending its neighbors, and adopted the Nanotechnology 2015 Action Plan that serves as a high-technology strategy for Germany's Nanotechnology Initiative.[12] The United States is also the world leader in corporate and venture capital investments in nanotechnology (Figure 6 and Figure 7). Global commercial nanotechnology spending is up 7 percent in 2012 to $9 billion while venture capital investment is down nearly 22 percent in 2010.[13]

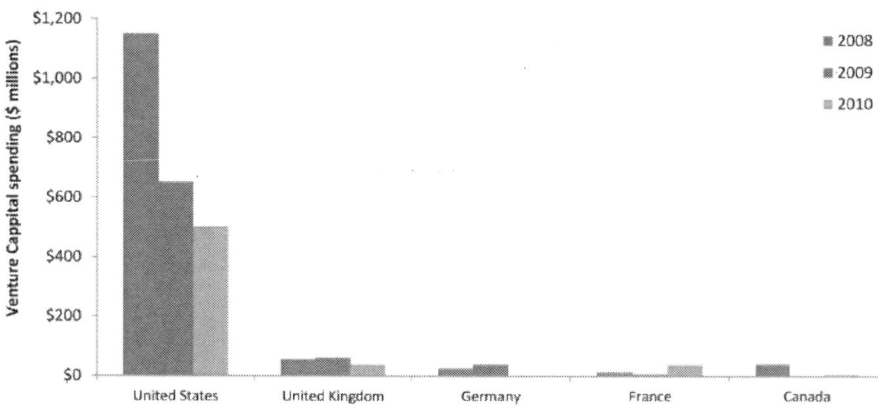

Source: A. Xue and D. Hwang, "Nanotechnology Funding: Corporations Gran the Reins," Lux Research, April 2011.

Note: No data were available for Germany's Venture Capital spending for 2010.

Figure 7. Venture capital spending for top five countries, 2008–2010.

RECOMMENDATIONS

In 2010, PCAST made recommendations (listed in Appendix E) to strengthen the role of the NNCO to act as a coordinating body for the NNI. While PCAST recognizes some improvement (described in the previous section, "Initiative Developments Since 2010") little progress has been made in the intervening time on several of these recommendations, which fall broadly into the following categories:

- *Strategic Planning:* the extent to which the overall NNI strategic plan has been effective at driving nanotechnology planning and resource allocation within the member agencies.
- *Program Management:* an assessment of additional actions that can be taken to strengthen the NNI leadership within the NSET and NNCO in areas of coordination and performance.
- *Metrics:* the extent to which metrics have been developed for measuring NNI outputs such as number of jobs created and the contribution of nanotechnology to the nation's economic growth.
- *Environmental, Health, and Safety Research:* an assessment of NNI progress toward understanding and managing the risks posed by nanotechnology to humans and the environment.

PCAST has determined that the NNI remains a successful cooperative venture. However, significant hurdles to an optimal structure and management of this broad initiative still persist. These hurdles include the level of authority that representatives appointed to NSET have within their home agencies to influence the budget allocations needed to meet NNI objectives, the inadequacy of mechanisms to solicit and act upon advice from outside of government, and the level of funding and capacity of the NNCO leadership to support the agencies in implementing programs that align with the NNI strategic plan. The lack of clear metrics for assessing the impacts of Federal investments in nanotechnology remains a concern, as do the still significant knowledge gaps that exist in the area of nanotechnologyrelated EHS impacts.

Strategic Planning Recommendations

In 2010, PCAST was concerned with a perceived deficiency in the agencies'strategic planning process regarding the management of and budgeting for nanotechnology research and development. In February 2011, the NSET Subcommittee produced an NNI strategic plan.'[4] PCAST views this document as evidence of progress toward achieving a more cohesive overall strategy for the initiative; however, individual agency contributions lack the cohesion of an overarching framework. Table 2 lists the three NNI agency documents containing a commonly understood definition of a strategic research plan.

Table 2. Agency-specific strategic research plans for nanotechnology

Agency	Title	Year
CDC	Strategic Plan for NIOSH Nanotechnology Research and Guidance	2009
EPA	Nanomaterial Research Strategy	2009
FDA	Nanotechnology Regulatory Science Research Plan	2011

Note: These documents contain strategic language for each agency in regard to nanotechnology.

The process for producing the strategic plan is still agency-driven, which could limit the broader NNI vision to those objectives that agency officials feel are sufficiently within reach. Also lacking is a clear connection between the goals and objectives of the NNI strategic plan with those of individual agencies. PCAST recommends taking the following actions:

NNCO in partnership with OSTP should work with the agencies to develop implementation plans for achieving the goals and objectives outlined in the 2011 NNI strategic plan.

Most participating agencies do not have strategic planning documents for nanotechnology. Agency officials assert that they have engaged in strategic planning activities to an extent that makes sense for their individual missions. Given that many of the agencies do not have a dedicated nanotechnology program but instead distribute their nanotechnology activities and budgetary outlays across multiple organizational units, it may be unnecessary—and perhaps infeasible—to create documented agency-specific strategic plans that could be aligned with the overarching NNI strategic plan. As an alternative to standalone strategic planning documents prepared by each participating agency, PCAST recommends the development of a nanotechnology implementation plan by each agency. Such plans should align with and communicate in detail how the agency will meet the objectives of the NNI strategic plan. Implementation plans could be discussed and revised in consultation with external stakeholders as well as with PCAST.

Ensure that senior agency officials capable of influencing funding decisions are participating fully and personally in strategic planning activities of the NSET. Officials at this level, in contrast to representatives active at the program or office level, could more effectively drive agency planning and budget allocations to meet NNI strategic directions.

Many members of the NSET are leaders in program management within their agencies. Some of these individuals, do not have direct influence over funding allocations, but they communicate NNI priorities back to the decision-makers in their agencies. Moreover, most agencies have no direct budget line dedicated to nanoscale science and engineering. Rather, each agency determines for itself how nano-technology is supported, by which programs and offices, and how resources are allocated relevant to its mission, often via an internal committee that reviews and suggests directions for the next budget year. This flexible arrangement acknowledges the importance of independence when planning agency activities but results in the suboptimal implementation of the broader NNI strategic plan. Ensuring that senior leaders, especially from the top six agencies of the NNI, engage directly and fully in the strategic planning process is one mechanism through which NNI objectives may gain the attention and commitment of those in a position to influence budget allocations. Under this arrangement, NSET representatives at the program level would continue to play a critical role in coordination, communication, and strategic planning activities.

The Nanotechnology Signature Initiatives should be fully supported in NNI budgets. To this end, we recommend that the Office of Management and Budget increase funding to these initiatives.

PCAST recognizes that the three existing NSIs play an enabling and foundational role in achieving both the objectives of the NNI and the larger strategic goal of global economic competitiveness over the next decade. As a result, during the next phase of strategic planning, agencies should consider the relevance of their nanotechnology investments to the existing NSIs and should commit the necessary resources to achieve these objectives. As these efforts continue, PCAST, through the recommendations that follow, re-emphasizes the importance of collecting and analyzing data about the outputs of NNI impacts broadly and the NSIs specifically so that their success can be measured in a transparent way.

The NSET Subcommittee should create Nanotechnology Signature Initiatives in other priority areas such as homeland security/national defense and human health.

The NSIs provide an effective way of focusing research across Federal agencies toward a common set of objectives to address a critical national need.

In addition to fully supporting the three existing NSIs, the NSET Subcommittee should consider creating new NSIs in areas of homeland security/national defense and human health research. While excellent work is being done already in these areas by one or more agencies, the NSIs would provide a framework for establishing a common set of objectives in these areas and facilitating more extensive interagency collaboration.

Program Management Recommendations

PCAST recognizes several actions taken by the NSET Subcommittee and NNCO since the publication of the 2010 NNI review. The identification of a dedicated NNCO staff person to coordinate nanotechnologyrelated EHS issues has already borne fruit in the publication of the 2011 NNI EHS Research Strategy. Likewise, there is now a position dedicated to liaising with industry and State stakeholders. To further these achievements, the NNCO should continue to broaden its impact and efficacy and improve its ability to coordinate and develop NNI programs and policies related to those programs. PCAST recommends OSTP take the following actions to facilitate these improvements:

Appoint the NNCO director as co-chair of the NSET Subcommittee.

The NNCO director currently is a non-voting member of the NSET Subcommittee. Appointing this person as co-chair, as is the case for another Presidential initiative, the Networking and Information Technology Research and Development (NITRD) Program, would enable the director to better coordinate with and support the agencies.

Change the requirement that the NNCO director must come from within the government to allow external, non-Federal experts the opportunity to direct the NNCO.

OSTP should recommend lifting the prohibition set forth in the 2001 Memorandum of Understanding (MOU) creating the NNCO that the NNCO director has to be a Federal employee. Enabling the NNCO in the future to engage talented leaders from outside of Federal service is one way to bring fresh perspectives and differing expertise into the management and coordination of the NNI. To be effective, such an individual must have a deep understanding of the cultures, missions, strategic priorities, and operations of the relevant Federal agencies.

Create a standing PCAST Nanotechnology Steering Committee of experts from industry, academia, and non-governmental organizations to provide more frequent and in-depth guidance to the overall initiative and to the signature initiatives.

Over the next decade, commercialization and translational activities will assume an increasingly important role in the success of the NNI. In contrast with support for upstream research, the locus for most of these activities lies outside of government. Therefore, advice and counsel from a broader set of non-Federal experts will be essential. PCAST recommends that a standing steering committee be created with experts from outside of government that meets regularly to provide more frequent external feedback to NSET and NNCO officials. This group could be modeled on the PCAST Advanced Manufacturing Partnership in which a PCAST member serves as an ex-officio member of an external group of leaders from industry and academia.'[5] The inclusion of members from non-governmental organizations will further the NNI goals of advancing responsible nanotechnology.

Dedicate 0.3 percent of NNI funding to the NNCO to ensure the appropriate staffing and budget to effectively develop, monitor, and assess NNI programs.

Currently, the NNCO receives 0.16 percent of each participating agency's allocated nanotechnology budget. The NNAP recommended in 2010 that the NNCO receive 0.3 percent of each participating agency's nanotechnology budget. The NSET Subcommittee discussed, but did not support, such an increase because current funding is based on programmatic needs rather than a fixed percentage of the NNI budget. PCAST's recommendations for restructuring the NNCO will require increased capacity and, consequently, an increase in its budget. These activities include management of and funding for the external advisory group recommended herein; stronger engagement with the business community, academia, and non-governmental organizations during the creation of the agencies'implementation plans; and enhanced coordination with state and local governments. Additionally, the NNCO may consider working with agencies such as the Bureau of Economic Analysis (BEA) or the Census Bureau to enumerate and collect the underlying data needed for the proposed workforce metrics (see metrics recommendation, below). It is reasonable for the NNCO budget to double in size over the next 3 years as it takes on more activities in support of the agencies.

Work with the NNCO director to develop a plan for increasing the NNCO budget in line with its new responsibilities.

OSTP should consider tasking the PCAST Nanotechnology Steering Committee recommended above as an advisory board and making management and budgetary discussions part of the group's deliberations. This would provide additional guidance to the NNCO regarding its administrative role and budgetary needs.

Metrics Recommendations

In 2010, PCAST recommended the development of measures that could be used to track the progress of nanotechnology outcomes, and PCAST reiterates the importance of metrics development here. To document progress made as a result of Federal investments in nanotechnology, it is essential that there is transparency and consistency in how each agency interprets the Federal definition of nanotechnology. As there is currently flexibility in the way individual agencies operationalize the definition, PCAST recommends taking the following actions to facilitate the characterization and quantification of the Federal nanotechnology portfolio of investments. Such actions will facilitate the collection of metrics that then can be used to estimate economic returns resulting from the NNI.

Agencies should develop a mission-appropriate definition of nanotechnology that enables the tracking of specific nanotechnology investments supported at the program level. The definition and funding details should be published in agency implementation plans to promote clarity.

This recommendation enables each agency to develop a mission-appropriate definition of nanotechnology to characterize its nanotechnology portfolio. Requiring each agency to publish its definition and the resulting budget allocations will improve clarity across the Federal nanotechnology portfolio and ensure that nanotechnology investments are accurately characterized.

The NNCO should track the development of metrics for quantifying the Federal nanotechnology portfolio and implement them to assess NNI outputs.

Current Federal efforts to measure public and private investment, scientific productivity, and workforce have been inconsistent and decentralized. The publication of agency-specific data will enable the NNCO to consistently track nanotechnology investments across the Federal government and enable it to report NNI impacts with greater confidence and transparency.

There is an extensive and growing body of high-quality academic research that is already working toward the establishment of nanotechnology metrics by drawing upon bibliometrics data from the public domain (e.g., publication and patent data).[16] Bibliometrics data are used as indicators of productivity beyond academia, often in the absence of other metrics from the private sector. As nanotechnology continues to mature and move closer toward commercialization, efforts to more accurately capture economic returns are picking up pace. Examples include the March 2012 International Symposium on Assessing Economic Impacts of Nanotechnologies sponsored jointly by the NNI and the Organization for Economic Co-Operation and Development held in Washington, DC, as well as the upcoming 2012 National Research Council review of the NNI.

A final area in need of metrics development is in the quantification of the nanotechnology workforce. Accurately categorizing agency-level nanotechnology investments will facilitate the identification of nanotechnology trainees, including the academic, scientific, and professional nanotechnology workforce for which there is currently a paucity of data.[17] One area where such tracking would have significant impact is in the identification of nanotechnology-related jobs for which there are no standard occupational codes. Good data on the workforce will enable the implementation of additional measures to identify and mitigate future threats to occupational health and safety.

PCAST recommends that NNCO serve as a central repository to collect these metrics and leverage advances in metrics-development to collect, track, and analyze data regarding publications, patents, educational activities, and the workforce to produce and publish its own statistics on behalf of the NSET. This undertaking is an integral component of cross-agency coordination of the Federal nanotechnology portfolio.

Environment, Health, and Safety Recommendations

PCAST acknowledges the significant progress made by the NNI to address potential environmental, health, and safety (EHS) risks of nanotechnology. Funding for nanotechnology-related EHS research has increased at a greater rate than the overall NNI budget, growing from $35 million in 2005 to $105 million in the 2013 request. This was appropriate, even necessary, to correct the significant imbalance present in 2005 between fundamental and applied research directed at new discoveries and the risk research that will help lower barriers to commercialization. PCAST is concerned that the 2013 request for support of nanotechnology-related EHS is only a modest increase over the 2012 request of $103 million. In light of the extensive list of outstanding research needs identified in both NNI's and NRC's recent EHS strategic planning documents,'[8] this is advisable only if significant improvements in coordination result in less redundancy among NNI agencies' nanotechnology-related EHS research portfolios and more effective multi-stakeholder, interagency, and international partnerships increase leverage of the Federal investments. The NSET should consider implementing the recommendation from the NRC EHS research strategy to direct $20–25 million toward informatics, partnerships, and instrumentation development without undercutting other research areas.

The NSET should establish high-level, cross-agency authoritative and accountable governance of Federal nanotechnology-related EHS research so that the knowledge created as a result of Federal investments can better inform policy makers.

PCAST acknowledges that the NSET has acted on our recommendation to identify a central coordinator for nanotechnology-related EHS research within NNCO. The EHS coordinator has done a laudable job developing and communicating the 2011 NNI EHS research strategy. However, there is still a lack of integration between nanotechnology-related EHS research funded through the NNI and the kind of information policy makers need to effectively manage potential risks from nanomaterials. The establishment of the Emerging Technologies Interagency Policy Coordination Committee (ETIPC) through OSTP has begun to bridge that gap, but without close integration between ETIPC and the NEHI working group, the gap may not be sufficiently narrowed. OSTP and the NSET Subcommittee should expand the charter of the NEHI working group to enable the group to address cross-agency nanotechnologyrelated policy issues more broadly.

The NSET should increase investment in cross-cutting areas of EHS that promote knowledge transfer such as informatics, partnerships, and instrumentation development.

The 2011 NNI EHS research strategy acknowledges the critical role that informatics, partnerships, and instrumentation development play in a comprehensive approach to addressing nanotechnology risks to human health and the environment. Nascent efforts in informatics should be supported so that advances can be accelerated in this critical cross-cutting area. Rather than continue to support the proliferation of databases that results from many new nano-EHS projects, the effort should be directed at enabling diverse communities to extract meaningful information from each other's work. New networks that connect researchers together, along with new tools for extracting information from Federally funded research, should be established and supported through the NNI. The findings of the December 2011 workshop to establish a Nanoinformatics 2020 Roadmap'[9] in conjunction with the 2011 NNI EHS research strategy can serve as a guide for new work in this area.

Significant progress has been made in the area of partnerships with numerous examples of multistakeholder and interagency collaboration underway. One of these is the *Nanorelease Project,*[2°] which brings together five NNI agencies, non-governmental organizations, a labor union, and several companies, among others, to develop methods for measuring the release of nanomaterials from commercial products. A specific area where better coordination could occur is in the area of occupational safety. The Occupational Safety and Health Administration (OSHA) should work with companies in a non-enforcement capacity to develop better tools for hazard communication similar to the National Institute of Occupational Health and Safety's (NIOSH) partnership program. This is especially important as the United States seeks to bring its hazard communication standard in alignment with the Globally Harmonized System of Classification and Labeling of Chemicals. Greater engagement by OSHA would also begin to address some of the difficulties companies face in implementing good health and safety programs in their nanomaterial workplaces. (See Appendix C.)

New modes of international cooperation, such as the joint funding of two environmental-impacts consortia by the EPA and the United Kingdom, have also emerged since the 2010 PCAST report. The NNI should increase funding for these cross-cutting activities to leverage the U.S. investment in nanotechnology-related EHS research.

CONCLUSION

PCAST continues to see extraordinary value in the NNI to support the President's overall innovation strategy, meet critical national needs, revitalize our economy, and serve as a platform for educating the next generation of American scientists. The United States continues to lead the world in government support for, and corporate and venture capital investments in, nanotechnology R&D. If properly managed, the NNI has the potential to support the development of products and services that will benefit society in multiple areas. However, challenges remain in successfully translating laboratory discoveries into the marketplace. Among these are continued concerns over the health and safety risks of nanomaterials and a cohesive strategy for commercialization.

Substantial progress has been made since the release of the 2010 assessment of the NNI. NSET's publication of the NNI strategic plan marks a significant step toward enhanced coordination and cooperation between NNI participating agencies. PCAST encourages participating agencies to develop their own strategic and implementation plans tailored to specific mission needs within the plan's overarching framework. Since 2010, the NNCO has been active in the management of the NNI, establishing new positions to coordinate EHS research and to strengthen relationships with state, local, and industrial partners and publishing the 2011 NNI EHS research strategy. PCAST recognizes significant overall progress in the EHS area, and recommends continued attention to this area as well as public engagement, particularly as products move to market. The three existing Nanotechnology Signature Initiatives have played an important enabling and foundational role in achieving both the objectives of the NNI and the larger strategic goal of global economic competitiveness over the next decade. PCAST recommends further support for these initiatives and suggests potentially establishing initiatives in the areas of homeland security/national defense and human health.

PCAST recognizes that the NNI has been successful to date but that a significant amount of work remains to fully realize the potential of the NNI. By continuing to apply the recommendations in the 2010 assessment and following those put forth in this document, PCAST feels that NNI will continue to be a success and that the benefits of investments in nanotechnology will translate to the public. Enhanced cooperation among the participating agencies through the development of specific implementation plans within the framework of the NNI strategic plan and through coordination with the NNCO will allow the agencies to address challenges and hurdles in a robust manner and strengthen the overall outcomes of the NNI.

APPENDIX A. EXPERTS PROVIDING INPUT TO PCAST

Sanjay Arora
Georgia Institute of Technology

Robert Celotta
National Institute of Standards and Technology

Yet-Ming Chiang
Massachusetts Institute of Technology

Cassandra Engemann
University of California at Santa Barbara

Linda Garlet
IDA Science and Technology Policy Institute

Piotr Grodzinski
National Institutes of Health

Barbara Herr Harthorn
University of California at Santa Barbara

Lori Henderson
National Institutes of Health

Patrick Herron
Duke University

Michael Holman
Lux Research

Evelyn Hu
Harvard University

Luciano Kay
Georgia Institute of Technology

Tim Lenoir
Duke University

Andrew Maynard
University of Michigan

Carlos Peña
Food and Drug Administration

Robert Pohanka
National Nanotechnology Coordination Office

Alan Porter
Georgia Institute of Technology

Mihail Roco
National Science Foundation

Nora Savage
Environmental Protection Agency

Andrew Schwartz
Department of Energy

Phil Shapira
Georgia Institute of Technology

Lewis Sloter
Department of Defense

Sally Tinkle
National Nanotechnology Coordination Office

Jan Youtie
Georgia Institute of Technology

APPENDIX B. NNI ORGANIZATIONAL CHART

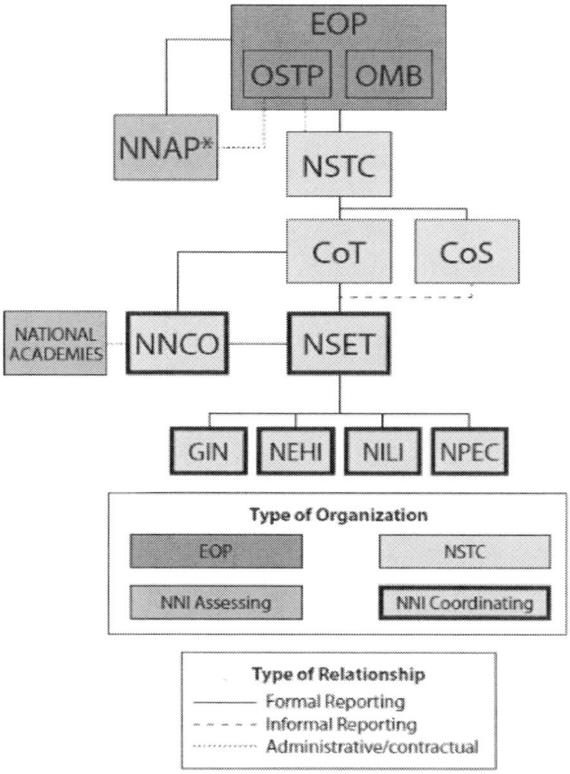

Source: National Science and Technology Council Committee on Technology Subcommittee on Nanoscale Science, Engineering, and Technology, "National Nanotechnology Initiative Strategic Plan," 2011, http:www.nano.gov/sites/default/files/pub_resource/2011_plan.pdf?q=nnistrategicplan2011.pdf.

*Executive Order 13349 designates the President's Council of Advisors on Science and Technology (PCAST) as the National Nanotechnology Advisory Panel (NNAP).

For acronym definitions within this chart, please see the Abbreviations Appendix at the end of the report.

Figure B-1. NNI organzational chart.

APPENDIX C . NANOTECHNOLOGY-RELATED ENVIRONMENT, HEALTH, AND SAFETY RESEARCH

As new modes of manufacturing are developed and explored, the need to address occupational health and safety issues will take on even greater urgency. Efforts to address workplace safety issues are limited by the lack of research, lack of rigorous information about the identity and demographics of the workforce, and by current practices and attitudes of employers towards workplace risk issues. An analysis of nanotechnology environmental, health, and safety (nano-EHS) research indicates that 5 percent of the collected nano-EHS papers published between 2001 and 2011 are of high relevance to workplace health and safety.[21] (See Figure C-1.) In addition, publications in the area of potential exposure to nanomaterials represent 12 percent of nano-EHS papers published in the last decade. A recent survey found that 59 percent of U.S. nanomaterials companies do not monitor the workplace for nanoparticles[22] despite government recommendations to do so.[23] This survey revealed a number of other attitudes and practices that demonstrate little progress since the publication of an earlier industry survey performed in 2007. It is therefore critical that the appropriate Federal agencies engage with companies in a non-regulatory capacity to increase their awareness of and ability to use the latest knowledge and guidance being generated on this topic.

The following figures demonstrate the growth in nano-EHS related research and the current practices and attitudes of employers toward workplace safety issues.

Figure C-1 shows the rate of publication of peer-reviewed nano-EHS research collected by the ICON Virtual Journal of Nano-EHS grew 18 percent between 2009 and 2001. Research of practical value to individuals charged with workplace safety, such as industrial hygienists, safety officers, and others (denoted by the category "Worker") lags a decade behind general nano-EHS research in terms of the number of publications per year.

In addition, there continues to be a large knowledge gap in the area of potential exposure to nanomaterials (denoted by the category "Exposure"); these publications represent only 12 percent of papers published in the last decade. These knowledge gaps persist despite increasing investments in EHS research. Figure C-2 shows cross-agency investment in nanotechnology EHS research between 2004 and 2012.

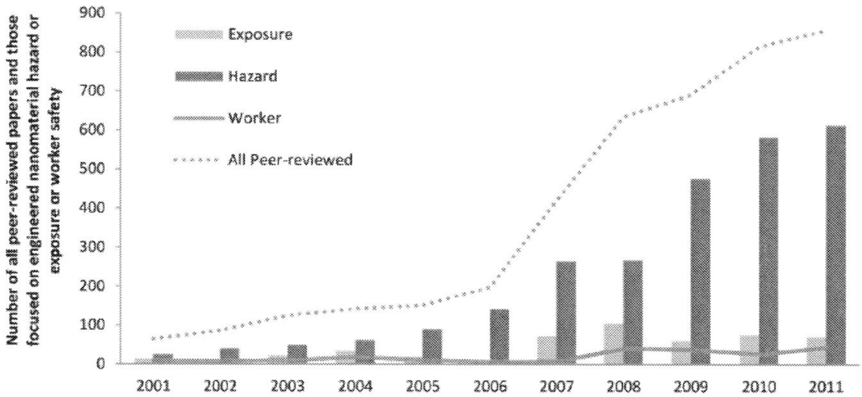

Source: International Council on Nanotechnology, Rice University, Nano-EHS Analysis Tool, http://icon.rice.edu/report.cfm.

Note: Data do not sum to the "All peer-reviewed" total because all categories not shown.

Figure C-1. Number of published, peer-reviewed papers listed in the ICON NanoEHS Virtual Journal.

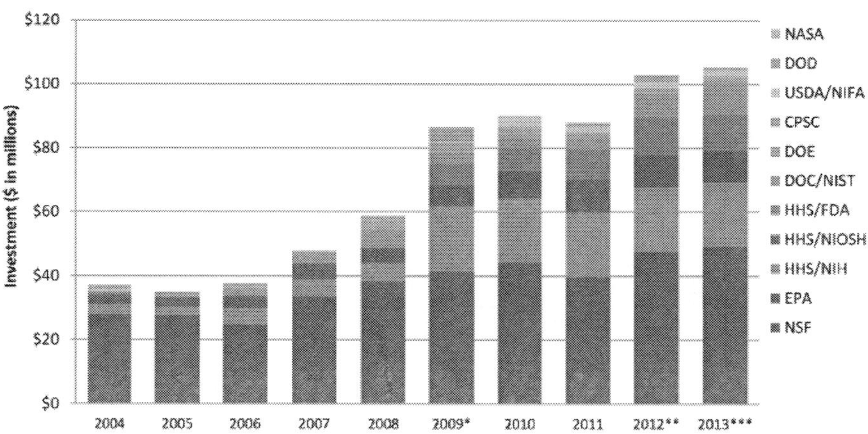

* Not including 2009 ARRA funding; ** 2012 figures are estimates; *** figures are proposed.

Figure C-2. Cross-agency investment in nanotechnology EHS research, 2004–2012.

Of greater concern is the inconsistency between attitudes of employers towards risk and the role of regulation in mitigating risk revealed by a recent survey of nanotechnology employers.[24] Depending on the material type, between 37 percent and 58 percent of employers believe (i.e., agree or strongly agree) that nanomaterials pose moderate to high risks to human health and/or the environment (Figure C-3).

While 84 percent believe that workplace safety should take priority over scientific and technological advances, 59 percent believe that waiting until safety studies are complete to commercialize nanotechnology will deprive society of too many potential benefits (Figure C-4).

Fifty-six percent believe that lack of information is an impediment in implementing nano-specific health and safety practices (Figure C-5) yet 77 percent believe that businesses are better informed about their own workplace safety needs than are government agencies.

As new knowledge of relevance to occupational health is being generated, OSHA should step up its efforts to work with companies to develop and implement responsible hazard communication and employee training programs using the most up-to-date information available.

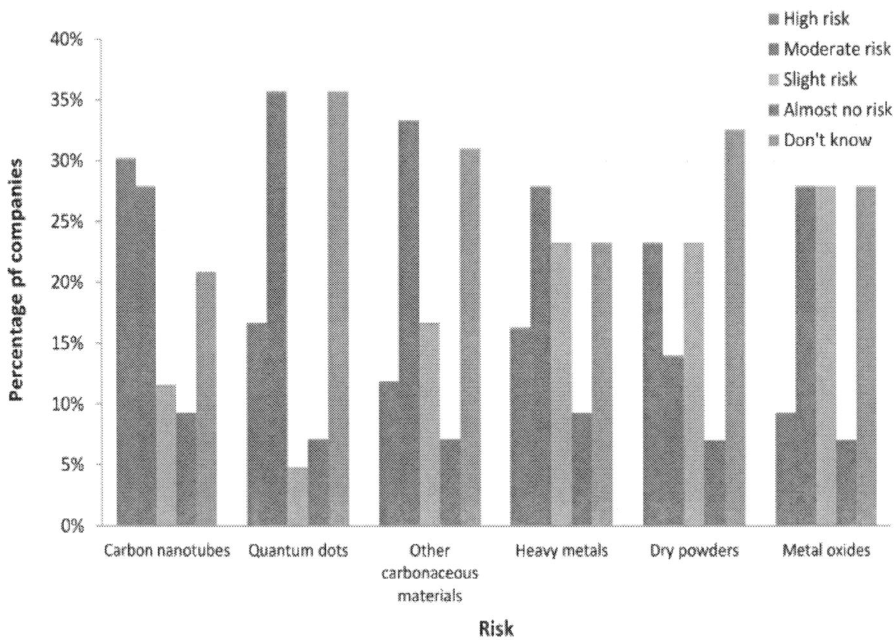

Figure C-3. Company views on risk across six types of nanomaterials.

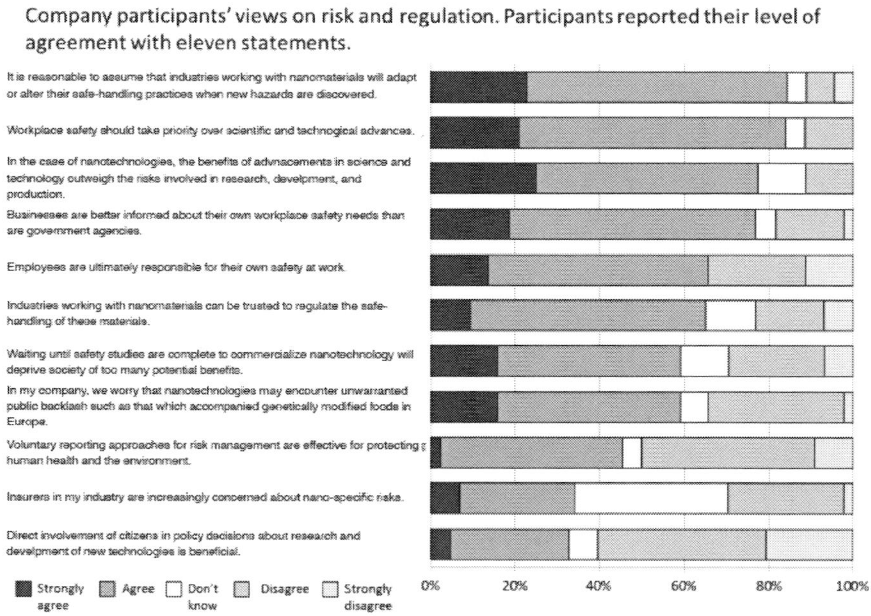

Figure C-4. Company attitudes toward risk and regulation.

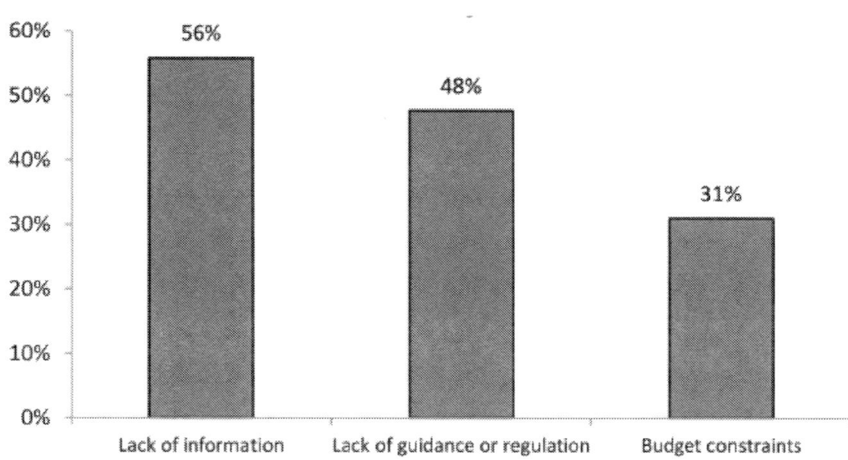

Figure C-5. Company-identified impediments to implementing nano-specific health and safety practices.

APPENDIX D. ADDITIONAL BIBLIOMETRICS

Figure D-1 illustrates the percent of citation counts by country for nanotechnology publications produced in 2009. As citations to any given paper typically rise over time, this figure captures the percentage-based global distribution of cited nanotechnology papers for one point in time. This figure shows that highest share of global citations are attributed to papers published with authors in the U.S. and EU-27.

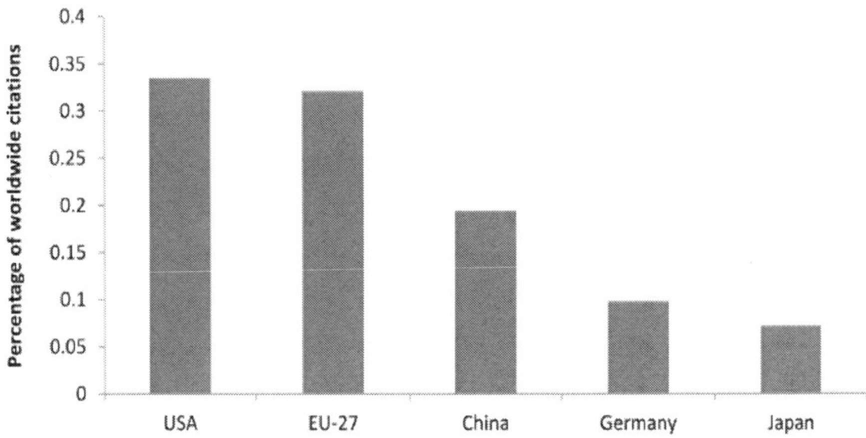

Source: L. Kay, P. Shapira, J. Youtie, and A. L. Porter, Program on Nanotechnology Research and Innovation System Assessment Georgia Institute of Technology, Atlanta, Georgia, with support from the National Science Foundation (NSF) through the Center for Nanotechnology in Society (Arizona State University; Award No. 0937591) [Data as of January 2010].

Notes: Search based on S. Arora, A. L. Porter, Y. Youtie, and S. Shapira, Capturing new develomnets in nanotechnology scientific output: A search strategy for publication records. STIP White Paper, Atlanta, georgia, 2011. This figure shows distribution for nanotechnology publications in 2009 as of 22 August 2011.

Figure D-1. Percentage by country of worldwide citations for nanotechnology articles in Web of Science (SCI, SSCI, and A&HCI).

Figures D-2 and D-3 illustrate citation and publication counts for selected journals. As one possible metric accounting for publication quality, three journals—*Science, Nature, and Proceedings ofthe National Academy of Sciences*—were selected; papers published in these journals are more often cited by other nanotechnology papers.

The U.S. and EU-27 exhibit a large publication lead in these journals compared to their Asian counterparts and hold a dominate position in terms of global publishing rate.

Due to a lag in uptake in usage, it is expected that the number of citation counts will continue to rise over time which helps to explain the appearance of a decline in citations after 2009 in Figure D-2.

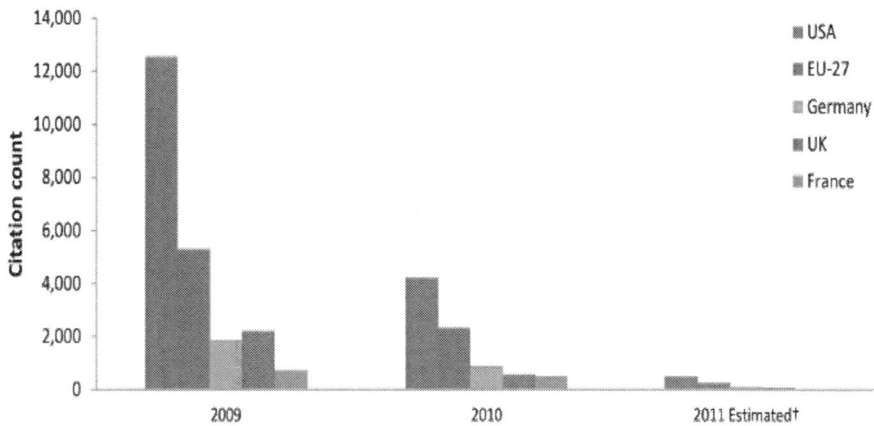

Source: L. Kay, P. Shapira, J. Youtie, and A. L. Porter, Program on Nanotechnology Research and Innovation System Assessment Georgia Institute of Technology, Atlanta, Georgia, with support from the National Science Foundation (NSF) through the Center for Nanotechnology in Society (Arizona State University; Award No. 0937591)

Notes: Selected journals include *Science, Nature, and Proceedings of the National Academy of Science.*

Search based on S. Arora, A. L. Porter, Y. Youtie, and S. Shapira, Capturing new develomnets in nanotechnology scientific output: A search strategy for publication records.

STIP White Paper, Atlanta, georgia, 2011.

† Annualized publication counts based on number of country-level publications identified as of 22 August 2011.

Additional 7% added to account for publications expected to be released in the next calendar year (i. e. in early 2012).

Figure D-2. Citation counts for nanotechnology articles from selected journals in Web of Science (SCI, SSCI, and A&HCI).

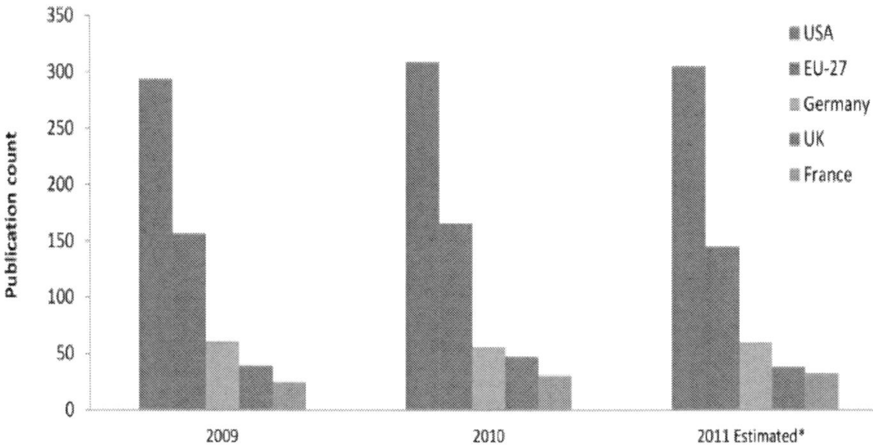

Source: L. Kay, P. Shapira, J. Youtie, and A. L. Porter, Program on Nanotechnology Research and Innovation System Assessment Georgia Institute of Technology, Atlanta, Georgia, with support from the National Science Foundation (NSF) through the Center for Nanotechnology in Society (Arizona State University; Award No. 0937591)

Notes: Selected journals include *Science, Nature, and Proceedings of the National Academy of Science.* Search based on S. Arora, A. L. Porter, Y. Youtie, and S. Shapira, Capturing new develomnets in nanotechnology scientific output: A search strategy for publication records.

STIP White Paper, Atlanta, georgia, 2011.

† Annualized publication counts based on number of country-level publications identified as of 22 August 2011. Additional 7% added to account for publications expected to be released in the next calendar year (i. e. in early 2012).

Figure D-3. Publication counts for nanotechnology articles from selected journals in Web of Science (SCI, SSCI, and A&HCI).

Figure D-4 ranks patent authorities by the number of priority patents filed in five year increments.

While nanotechnology patent filings grew in all patent authorities over the 20 year time span, U.S. filings grew significantly after the year 2000, eventually surpassing those filed in the International Patent System.

The proliferation of patent filings in the United States could indicate a shifting trend in corporate filing strategies based upon factors such as filing fees, market, and/or competitive locations.

Figure D5 highlights the ration of corporate nanotechnology patent filings to publications and shows a perceptible shift over time,[25] indicating a possible change in corporate emphasis from nanotechnology discovery to application in the early 2000s.

U.S. companies exhibit a smaller ratio of patent filingstopublications26 than do non-U.S. companies, possibly indicating that U.S. corporations tend to publish more articles on average compared with the generation of patent applications than do non-U.S. companies.

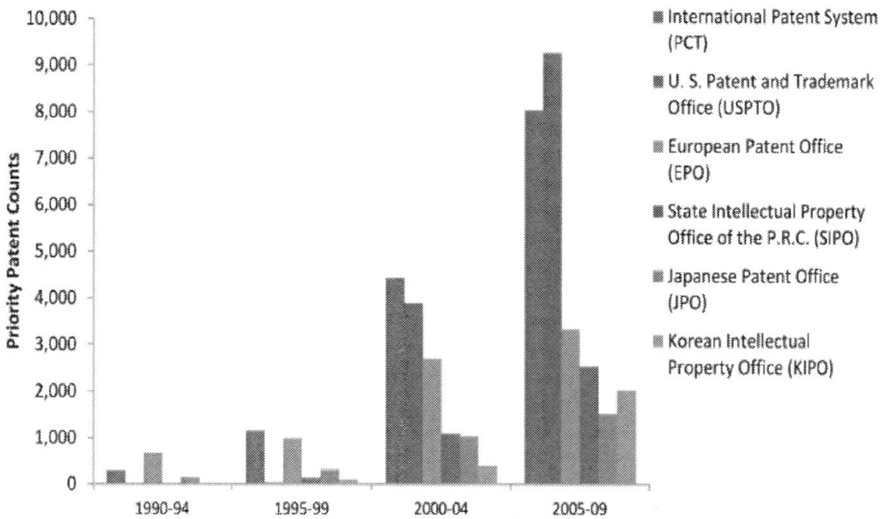

Source: L. Kay, P. Shapira, J. Youtie, and A. L. Porter, Program on Nanotechnology Research and Innovation System Assessment Georgia Institute of Technology, Atlanta, Georgia, with support from the National Science Foundation (NSF) through the Center for Nanotechnology in Society (Arizona State University; Award No. 0937591) [Data as of January 2010]

Notes: Based on "nanotechnology" definition by A. L. Porter, J. Youtie, P. Shapira, and D. Schoeneck, "Refining Search terms for Nanotechnology," *Journal of Nanoparticle Research,* 10 (5): 715-728.

Proirity applications are patent documents in PATSTAT that report priority sequence of "1".

Figure D-4. Counts of priority patent applications in nanotechnology in PATSTAT, by year and patent authority, 1990–2009.

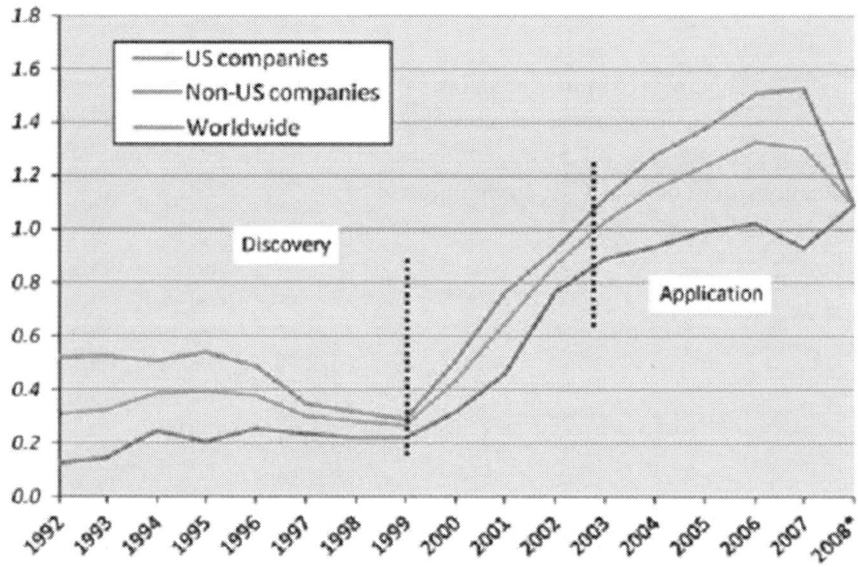

Source: P. Shapira, J. Youtie, and L. Kay (2011) "National Innovation Systems and the Globalization of Nanotechnology Innovation," *Journal of technology Transfer* 36: 587-604.

Note: At the time data were collected, 2008 was an incomplete year. As as result, data were annualized for number of worldwide patents, which produces a single ratio.

Figure D-5. Ratio of corporate nanotechnology patent applications to publications, 1992–2008.

APPENDIX E. 2010 PCAST RECOMMENDATIONS

Program Management

Recommendation 2-1: Strengthen the NNCO

The NNCO should broaden its impact and efficacy and improve its ability to coordinate and develop NNI programs and policies related to those programs. OSTP should facilitate these improvements by taking the following actions:

- Require each agency in the NNI to have senior representatives with decision-making authority participate in coordination activities of the NNI.

- Strengthen the NNCO to enhance its ability to act as the coordinating entity for the NNI.
- Mandate that the NNCO develop metrics for program outputs and that it works with the Bureau of Economic Analysis to develop metrics and to collect data on the economic impacts of the NNI.
- Appoint two individuals to the NNCO to lead interagency coordination of efforts in the areas of EHS research and standards development, respectively. Dedicate 0.3 percent of NNI funding to the NNCO to ensure the appropriate staffing and budget to effectively develop, monitor and assess NNI programs.

Recommendation 2-2: Focus on Commercialization

In a budget planning process coordinated by OSTP, each agency should continually re-evaluate its NNI balance of investments among the PCAs, with an enhanced focus on commercialization, which would include maintaining the current level of investment in research and doubling the investment in nanomanufacturing (PCA5) over the next five years.

Recommendation 2-3: Signature Initiatives

Each Signature Initiative's lead agency should develop coordinated milestones, promote strong educational components, and create public-private partnerships to leverage the outcomes of the initiatives.

Each lead agency also should develop strategies for monitoring, evaluating, and disseminating outcomes.

Recommendation 2-4: Education

The agencies of the NNI should continue making investments in innovative and effective education, and the NNCO should consider commissioning a comprehensive evaluation of the outcomes of the overall investment in NNI education.

Recommendation 2-5: Societal Impacts

The NSET Subcommittee should develop a clear expectation and strategy for programs in the societal dimensions of nanotechnology.

An effective program in societal implications would have well-defined areas of focus, clearly articulated outcomes as well as plans for assessing and evaluating those outcomes, and partnerships that leverage the value of its activities.

Ultimately, the inclusion of such programs in the NNI has the goal of streamlining nanotechnology innovation and its positive impact on society, and the creation of new jobs, opportunities and a robust economy.

Outputs

Recommendation 3-1: Nanomanufacturing and Commercialization

The NSF, DOE, Department of DOD, NIST, and NIH should include a greater emphasis on manufacturing, and commercialization while maintaining or expanding the level of basic research funding in nano-technology. Specifically, over the next five years, the Federal Government should double the funding devoted to nanomanufacturing (PCA5).

In addition, the Federal Government should launch at least five government-industry-university partnerships, using the Nanoelectronics Research Initiative as a model.

The Federal Government should also support at least five Signature Initiatives over the next two to three years, with each Signature Initiative funded at levels adequate to achieve its stated goals, presumably between $20 million and $40 million annually.

Recommendation 3-2: Job Creation

The Department of Commerce and the Small Business Administration should advise the NNI on how to ensure that its programs create new jobs in the United States, including coordinating with State efforts, and economic impact should be an explicit metric in the second decade of the NNI.

Recommendation 3-3: Workforce Retention

Congress and the Administration need to take steps to retain scientific and engineering talent trained in the United States by developing a program to provide U.S. Permanent Resident Cards for foreign individuals who receive an advanced degree in science or engineering at an accredited institution in the United States and for whom proof of permanent employment in that scientific or engineering discipline exists.

Recommendation 3-4: Moving Nanotechnology to Market

The DOE, DOD, NIST, NIH, NCI, FDA, and NIST should clarify the development pathway and increase their emphasis on transitioning nanotechnology to commercialization, including making sustained meaningful investments in focused areas to help accelerate technology transfer to the marketplace.

Environment, Health, and Safety

Recommendation 4-1: Risk Identification

The NSET Subcommittee's NEHI working group should develop clear principles to support the identification of plausible risks associated with the products of nanotechnology.

Recommendation 4-2: Strategic Planning

The NSET Subcommittee's NEHI working group should further develop and implement a crossagency strategic plan that links EHS research activities with knowledge gaps and decision-making needs within government and industry.

Recommendation 4-3: Organizational Changes

The NSET Subcommittee and OSTP should foster administrative changes and communications mechanisms that will enable the NNI to better embrace the EHS issues associated with nanotechnology research, development, and commercialization.

- The NSET Subcommittee co-chairs should assign an individual to NNCO to oversee interagency efforts that address nanotechnology EHS.
- OSTP and the NSET Subcommittee should expand the charter of the NEHI working group to enable the group to address cross-agency nanotechnology-related policy issues more broadly.
- The NSET Subcommittee should explore mechanisms that enable the NEHI working group to more effectively receive input and advice from nongovernment experts in the field of emergent risks.

Recommendation 4-4: Information Resources

The NSET Subcommittee's NEHI working group should develop information resources on crosscutting nanotechnology EHS issues that are relevant to businesses, health and safety professionals, researchers, and consumers.

ABBREVIATIONS

A&HCI	Arts and Humanities Citation Index
ARPAE	Advanced Research Projects Agency for Energy
ARRA	American Recovery and Reinvestment Act of 2009
ATE	Advanced Technology Education
DOC	Department of Commerce
DOD	Department of Defense
DOE	Department of Energy
EERE	Energy Efficiency and Renewable Energy
EHS	Environment, Health, and Safety
ENM	Engineered Nanomaterials
EPA	Environmental Protection Agency
EU	European Union
FY	Fiscal Year
HHS	Department of Health and Human Services
ICON	International Council on Nanotechnology
ISL	Industry and State Liaison
MOU	Memorandum of Understanding
NACK	National Nanotechnology Applications and Career Knowledge
NASA	National Aeronautics and Space Administration
NEHI	Nanotechnology Environmental and Health Implications
NIH	National Institutes of Health
NILI	Nanomanufacturing, Industry Liaison, and Innovation
NIOSH	National Institute for Occupational Safety and Health
NIST	National Institute of Standards and Technology
NITRD	Networking and Information Technology Research and Development
NNAP	National Nanotechnology Advisory Panel
NNCO	National Nanotechnology Coordination Office
NNI	National Nanotechnology Initiative
NSEC	Nanoscale Science and Engineering Centers
NSET	National Science, Engineering, and Technology (Subcommittee)
NSI	Nanotechnology Signature Initiative
NSF	National Science Foundation
NSTC	National Science and Technology Council
OMB	Office of Management and Budget

OSHA	Occupational Health and Safety Administration
OSTP	Office of Science and Technology Policy
PATSTAT	Patent Statistical Database
PCA	Program Component Area
PCAST	President's Council of Advisors on Science and Technology
R&D	Research and Development
RUSNANO	Russian Nanotech Corporation
SCI	Science Citation Index
SSCI	Social Science Citation Index

End Notes

[1] "NNI Vision, Goals, and Objectives," National Nanotechnology Initiative website, http://www.nano.gov/about-nni/what/vision-goals.

[2] Bill Clinton, "President Clinton's address to the California Institute of Technology," speech, California Institute of Technology, Pasadena, CA, January 21, 2000, http://today.caltech.edu/theater/list?subset=culture&story_count=end.

[3] "A Strategy for American Innovation: Driving towards Sustainable Growth and Quality Jobs," White House website, http://www.whitehouse.gov/administration/eop/nec/StrategyforAmericanInnovation.

[4] "Nanotechnology Signature Initiatives," National Nanotechnology Initiative website, http://www.nano.gov/html/research/signature_initiatives.html.

[5] "National Nanotechnology Initiative," Supplement to the President's FY 2012 Budget, http://www.nano.gov/sites/default/files/pub_resource/nni_2012_budget_supplement.pdf.

[6] Based on definition of "nanotechnology" by A. L. Porter, J. Youtie, P. Shapira, and D. Schoeneck, "Refining Search Terms for Nanotechnology," *Journal of Nanoparticle Research*, 10 (5): 715–728.

[7] Executive Office of the President, President's Council of Advisors on Science and Technology, "Report to the President and Congress on the Third Assessment of the National Nanotechnology Initiative," March 12, 2010, p. vi.

[8] National Science and Technology Council Committee on Technology Subcommittee on Nanoscale Science, Engineering, and Technology, "NNI Environmental, Health, and Safety (EHS) Research Strategy," 2011, http://www.nano.gov/sites/default/files/pub_resource/nni_2011_ehs_research_strategy.pdf.

[9] National Research Council, Committee to Develop a Research Strategy for Environmental, Health, and Safety Aspects of Engineered Nanomaterials, "A Research Strategy for Environmental, Health, and Safety (EHS) Aspects of Engineered Nanomaterials (ENMs)," 2012, http://www.nap.edu/catalog.php?record_id=13347.

[10] The centers are NSEC for Integrated Nanopatterning and Detection Technologies; NSEC for Nanoscale Systems in Information Technologies; NSEC for Science of Nanoscale Systems and their Device Applications; Center for Electronic Transport in Molecular Nanostructures; Center for Biological and Environmental Nanotechnology; and NSEC for the Directed Assembly of Nanostructures.

[11] J. Rogers, J. Youtie, A. Porter, and P. Shapira, "Assessment of Fifteen Nanotechnology Science and Engineering Centers' (NSECs) Outcomes and Impacts: Their contribution to NNI Objectives and Goals," http://www.nsf.gov/crssprgm/ nano/reports/Assessment_ 2011+May+12+of+NSEC+by+GaTech_FinalReport_56p_web.pdf.

[12] Federal Ministry of Education and Research (2011). Action Plan Nanotechnology 2015, http://www.bmbf.de/pub/akionsplan_nanotechnologie_2015_en.pdf.

[13] See also, M. Roco, "Nanotechnology: From Discovery to Innovation and Socioeconomic Projects," *Chemical Engineering Progress* May 2011, specifically Table 1, "Nanotechnology development can be characterized by a variety of indicators."

[14] National Science and Technology Council Committee on Technology, Subcommittee on Nanoscale Science, Engineering, and Technology, "National Nanotechnology Initiative Strategic Plan," 2011, http://www.nano.gov/sites/default/files/pub_resource/2011_ strategic _plan.pdf.

[15] About the Advanced Manufacturing Partnerships: http://www.whitehouse.gov/administration/ eop/ostp/pcast/amp.

[16] *See* J. Youtie, P. Shapira, and A. L. Porter, "Nanotechnology Publications and Citations by Leading Countries and Blocs," *Journal of Nanoparticle Research* 10 (2008): 981–986 ; J. Wang and P. Shapira, "Funding Acknowledgement Analysis—An Enhanced Tool to Investigate Research Sponsorship Impacts: The Case of Nanotechnology," *Scientometics* 87 (3): 563–586; L. Leydesdorff and I. Rafols "The Local Emergence and Global Diffusion of Research Technologies: An Exploration of Patterns of Network Formation" *Journal of the American Society for Information Science and Technology* 62 (5): 846–860; C. Huang and Y. Wu, "State-Led Technological Development: A Case of China's Nanotechnology Development," *World Development*, forthcoming.

[17] J. Walsh and C. Ridge, "Knowledge Production and Nanotechnology: Characterizing American Dissertation Research, 1999-2009," *Technology in Society*, in press.

[18] National Science and Technology Council Committee on Technology Subcommittee on Nanoscale Science, Engineering, and Technology, "NNI Environmental, Health, and Safety (EHS) Research Strategy," 2011, http://www.nano.gov/sites/default/files/pub_resource/ nni_2011_ehs_research_strategy.pdf; and Committee to Develop a Research Strategy for Environmental, Health, and Safety Aspects of Engineered Nanomaterials; National Research Council, "A Research Strategy for Environmental, Health, and Safety (EHS) Aspects of Engineered Nanomaterials (ENMs)," 2012, http://www.nap.edu/catalog.php? record_id=13347.

[19] About the Nanoinformatics 2011 workshop: http://www.nanotechinformatics.org/.

[20] About the Nanorelease Project: http://www.ilsi.org/ResearchFoundation/Pages/ NanoRelease Overview.aspx.

[21] International Council on Nanotechnology (ICON), Virtual Journal of NanoEHS, http://icon.rice.edu/virtualjournal.cfm.

[22] C. D. Engeman, L. Baumgartner, B. M. Carr, A. M. Fish, J. Meyerhofer, T. A. Satterfield, P. A. Holden, and B. H. Harthorn, "Governance Implications of Nanomaterials Companies' Inconsistent Risk Perceptions and Safety Practices," *Journal of Nanoparticle Research* 14: 749. doi 10.1007/s11051-012-0749-0.

[23] Department of Health and Human Services, National Institute for Occupational Safety and Health, "Approaches to Safe Nanotechnology, Managing the Health and Safety Concerns Associated with Engineered Nanomaterials," Publication No. 2009–125, March 2009.

[24] C. D. Engeman, L. Baumgartner, B. M. Carr, A. M. Fish, J. Meyerhofer, T. A. Satterfield, P. A. Holden, and B. H. Harthorn, "Governance Implications of Nanomaterials Companies'

Inconsistent Risk Perceptions and Safety Practices," *Journal of Nanoparticle Research*. 14: 749. doi 10.1007/s11051-012-0749-0.

[25] P. Shapira, J. Youtie, and L. Kay, "National Innovation Systems and the Globalization of Nanotechnology Innovation," *Journal of Technology Transfer* 36: 587–604.

[26] In Figure D-5, the Y-axis provides a ratio of corporate nanotechnology patent applications to corporate nanotechnology publications by year. The ratio is based on patent applications in worldwide patent offices and is then separated out by U.S. and foreign companies. Data were collected based on patent families, which minimizes the potential for technology duplication or double-counting.

In: Nanotechnology
Editors: S.M. Goldfarb, H. Morgan
ISBN: 978-1-62257-596-1
© 2013 Nova Science Publishers, Inc.

Chapter 2

NANOTECHNOLOGY: IMPROVED PERFORMANCE INFORMATION NEEDED FOR ENVIRONMENTAL, HEALTH, AND SAFETY RESEARCH*

United States Government Accountability Office

WHY GAO DID THIS STUDY

Nanotechnology involves the ability to control matter at approximately 1 to 100 nanometers. Worldwide trends suggest that products that rely on nanotechnology will be a $3 trillion market by 2020. However, some of the EHS impacts of nanotechnology are unknown. The NSTC coordinates and oversees the NNI, an interagency program that, among other things, develops national strategy documents for federal efforts in nanotechnology.

In this context, GAO examined: (1) changes in federal funding for nanotechnology EHS research from fiscal years 2006 to 2010; (2) the nanomaterials that NNI member agencies' EHS research focused on in fiscal year 2010; (3) the extent to which NNI member agencies collaborate with stakeholders on this research and related strategies; and (4) the extent to which NNI strategy documents address desirable characteristics of national

* This is an edited, reformatted and augmented version of United States Government Accountability Office, Publication No. GAO-12-427, dated May 2012.

strategies. GAO's review included seven NNI agencies that funded 93 percent of the EHS research dollars in fiscal year 2010. This report is based on analysis of NNI and agency documents and responses to a questionnaire of nonfederal stakeholders.

WHAT GAO RECOMMENDS

GAO recommends that the Director of the Office of Science and Technology Policy (OSTP), which administers the NSTC, (1) coordinate development of performance information for NNI EHS research needs and publicly report this information; and (2) estimate the costs and resources necessary to meet the research needs. OSTP and the seven included agencies neither agreed nor disagreed with the recommendations.

WHAT GAO FOUND

From fiscal years 2006 to 2010, the National Science and Technology Council (NSTC) reported more than a doubling of National Nanotechnology Initiative (NNI) member agencies' funding for nanotechnology environmental, health, and safety (EHS) research—from approximately $38 million to $90 million. Reported EHS research funding also rose as a percentage of total nanotechnology funding over the same period, ending at about 5 percent in 2010. However, GAO identified several reporting problems that raise concerns about the quality of EHS funding data reported. For example, for 18 percent of the 2010 projects GAO reviewed that were reported as EHS research, it was not clear that the projects were primarily directed at EHS risks. In addition, NNI member agencies did not always report funding using comparable data. The absence of detailed guidance on how agencies should report funding for their nanotechnology research has contributed to these problems, as GAO also reported in 2008 and made a related recommendation.

In 2010, EHS research at the NNI member agencies GAO reviewed most frequently focused on carbon nanotubes, nanosilver, and nanoscale titanium dioxide. NNI has not prioritized nanomaterials for EHS research, but NNI's 2011 EHS research strategy outlines criteria for NNI member agencies to use in doing so. It is too soon to tell how these criteria will influence NNI member agencies' decisions about which nanomaterials to prioritize, and it is unclear if information needed to use the NNI criteria is available.

The NNI member agencies have collaborated extensively on EHS research and strategies. They have collaborated through the NSTC to develop joint EHS research strategies and have initiated numerous formal collaborative EHS research projects. Nonfederal stakeholders who responded to GAO's web-based questionnaire on nanotechnology EHS research told GAO that they benefited from collaboration with the NNI member agencies but identified some challenges, including a lack of funding and limited awareness of collaboration opportunities, among others. Most respondents rated the 2011 NNI EHS research strategy as somewhat or very effective at addressing nanotechnology EHS research needs.

NNI strategy documents for EHS research issued by the NSTC address two and partially address the other four of the six desirable characteristics of national strategies identified by GAO that offer a management tool to help ensure accountability and more effective results. For example, the NNI strategy documents provide a clear statement of purpose, define key terms, and discuss the quality of currently available data, among others. However, they do not include performance information—such as performance measures, targets, and time frames for meeting those measures—that would allow stakeholders to evaluate progress towards the goals and research needs of the NNI. In addition, the documents do not include, or sufficiently describe, estimates of the costs and resources needed for the strategy. Without this information, it may be difficult for agencies and stakeholders to implement the strategy and report on progress toward achieving the research needs and assess if investments are commensurate with costs of the identified needs.

ABBREVIATIONS

CPSC	Consumer Product Safety Commission
DOD	Department of Defense
EHS	environmental, health, and safety
EPA	Environmental Protection Agency
FDA	Food and Drug Administration
IANH	International Alliance for NanoEHS Harmonization
ICON	International Council on Nanotechnology
NCI	National Cancer Institute
NEHI	Nanotechnology Environmental and Health Implications

NGO	nongovernmental organizations
NIEHS	National Institute of Environmental Health and Science
NIH	National Institutes of Health
NIOSH	National Institute for Occupational Safety and Health
NIST	National Institute of Standards and Technology
NNCO	National Nanotechnology Coordination Office
NNI	National Nanotechnology Initiative
NRC	National Research Council
NSET	Nanoscale Science, Engineering, and Technology
NSF	National Science Foundation
NSTC	National Science and Technology Council
OMB	Office of Management and Budget
OSTP	Office of Science and Technology Policy
PCA	program component area
PCAST	President's Council of Advisors on Science and Technology
R&D	research and development Recovery Act American Recovery and Reinvestment Act of 2009
SAIC	Science Applications International Corporation
UK	United Kingdom

May 21, 2012
The Honorable Barbara Boxer
Chairman
Committee on Environment and Public Works
United States Senate

Dear Madam Chairman:

Nanotechnology is the understanding and control of matter at approximately 1 and 100 nanometers, known as the nanoscale. For illustration, a sheet of paper is about 100,000 nanometers thick, a human hair is about 80,000 nanometers in diameter, and three gold atoms lying side by side are about 1 nanometer long. Unusual properties can emerge in materials manufactured at the nanoscale—including electrical, magnetic, mechanical, optical, and thermal properties—that differ in important ways from the properties of conventionally scaled materials. Materials at this scale that have been manufactured are sometimes referred to as engineered nanomaterials.[1] Nanotechnology has enabled or facilitated novel research in areas such as computing, medicine, energy conversion and storage, water purification, agriculture and food systems, synthetic biology, aerospace, and geoengineering. Consumer products that incorporate nanotechnology are as diverse as clothing, cosmetics, household appliances, and sporting goods. For example, nanoscale particles of titanium dioxide used in sunscreens act as physical filters that absorb UV light without affecting the opacity of the product. A 2010 study estimated that values for products enabled by nanotechnology were worth about $91 billion in the United States and $254 billion worldwide in 2009.[2] Trends suggest that the number of nanotechnology products and workers employed in related fields will double every 3 years worldwide, achieving a $3 trillion market and 6 million workers by 2020.

Even with its increasing commercialization, much remains unknown about nanotechnology, including some of the environmental, health, and safety (EHS) impacts of nanomaterials. For example, there are few tools and methods—such as models to predict the behavior of nanomaterials in the environment—for conducting research. It is difficult to assess the risk of nanomaterials because these materials are too varied to generalize how they will behave; nonetheless, risks associated with particular uses of specific nanomaterials can be assessed. Further, there is little information on the number of workers exposed to nanomaterials in the workplace or the effects on human health of such exposure. However, as we reported in 2010, some research indicates that the toxicity of certain nanomaterials, such as some forms of carbon nanotubes and nanoscale titanium dioxide, may pose a risk to human health.[3]

The National Nanotechnology Initiative (NNI) is a federal interagency program that seeks to expedite the discovery, development, and deployment of nanoscale science, engineering, and technology to serve the public good through coordinated research and development aligned with the missions of its member agencies.[4] The NNI informs and influences the federal nanotechnology budget and planning process through its member agencies and through the National Science and Technology Council (NSTC), an entity administered by the Office of Science and Technology Policy (OSTP) that is the principal means by which the executive branch coordinates science and technology policy. The NSTC's Committee on Technology coordinates the NNI under the Nanoscale Science, Engineering, and Technology (NSET) Subcommittee. The National Nanotechnology Coordination Office (NNCO), which reports to the Committee on Technology, provides technical and administrative support to the NSET. The NNI, through the NNCO and appropriate agencies, among others, is directed to (1) establish goals, priorities, and metrics for evaluation for federal nanotechnology research, development, and other activities; (2) invest in federal research and development programs in nanotechnology and related sciences to achieve these goals; and (3) provide for interagency coordination of federal nanotechnology research, development, and other activities.

The federal nanotechnology commitment is significant. Cumulative reported NNI funding since fiscal year 2001 amounts to more than $14 billion. Nanotechnology research projects and activities undertaken by the NNI member agencies are categorized into eight program component areas.[5] Agencies report their research funding for each program component area to the Office of Management and Budget (OMB) as part of the annual federal budget process. The seventh program component area comprises research "primarily directed at understanding the EHS impacts of nanotechnology development and corresponding risk assessment, risk management, and methods for risk mitigation," which we refer to in this report as EHS research.

The NSTC issued three key NNI strategy documents in 2011 that, as a whole, seek to establish shared goals and research needs; describe NNI member agencies' activities; and provide guidance for agency leaders, program managers, and the research community regarding planning and implementation of nanotechnology research and development investments and activities, including those related to EHS research. 6 Jointly, these documents form a national strategy for federal nanotechnology efforts—such strategies tend to cut across sectors of the economy and levels of government and involve a range of public and private organizations and entities.[7] The 2011 NNI strategic plan provides a vision and overarching goals for the program, including responsible development of nanotechnology, which includes EHS research. The 2011 NNI EHS research strategy expands on the goal of the responsible development of nanotechnology by describing the state of science and research needed to ensure that nanotechnology provides maximum benefits to the environment and human well-being. The NNI Supplement to the President's 2012 Budget serves as the annual report for the NNI and describes recent activities by the NNI member agencies and funding of the NNI, including the funding level for the program component area related to EHS research. These NNI strategy documents are the principal public record of the NSTC's coordinating activities and reflect the NSTC's guidance to agencies as they separately determine their budgets and activities for nanotechnology EHS research.

In prior work, we have identified six desirable characteristics for national strategies and reviewed strategies that the federal government has used to plan for and report on crosscutting issues including emergency preparedness, financial literacy, influenza, and terrorism for these characteristics. 8 These desirable characteristics help shape the policies, programs, priorities, resource allocations, and standards that would enable federal agencies and other stakeholders to implement the strategies and achieve the identified results. National strategies that address these characteristics offer policymakers and implementing agencies a management tool that can help ensure accountability and more effective results.

In this context, you asked us to review federal nanotechnology EHS research. This report examines (1) changes in federal funding for nanotechnology EHS research from fiscal years 2006 to 2010; (2) the nanomaterials that NNI member agencies focused on in their EHS research in fiscal year 2010; (3) the extent to which NNI member agencies collaborate with stakeholders on nanotechnology EHS research and related strategies; and (4) the extent to which NNI strategy documents address desirable characteristics of national strategies.

To conduct this work, we reviewed EHS research efforts funded by seven NNI agencies, which collectively funded 93 percent of EHS research dollars in fiscal year 2010: the National Science Foundation (NSF), National Institutes of Health (NIH), Environmental Protection Agency (EPA), National Institute for Occupational Safety and Health (NIOSH), Food and Drug Administration (FDA), National Institute of Standards and Technology (NIST), and the Consumer Product Safety Commission (CPSC). The first six of these agencies represent the top six providers of EHS research funding from fiscal year 2006 to fiscal year 2010, and CPSC has an important role in ensuring the safe use of nanotechnology in consumer products.

To examine recent changes in federal funding for nanotechnology EHS research, we reviewed information published by NSTC in the NNI Supplements to the President's Budget. Specifically, we reviewed funding reported in each program component area for fiscal years 2006 through 2010 for all NNI member agencies funding nanotechnology research.[9] For the dollar amounts that we adjusted for inflation, we used the Biomedical Research and Development Price Index to report funding in constant 2010 dollars.[10] We consulted with agency and OMB officials to determine the type of budget information reported in these documents for the agencies' investments. For fiscal year 2010, the most recent year for which agencies' nanotechnology investment data were available, we also collected quantitative and qualitative project-level data on all research projects that the seven selected agencies had categorized as EHS research (program component area 7). The agencies do not report project-level data to OMB annually but report their total EHS research funding to OMB annually for inclusion in the NNI budget supplements published by NSTC. Therefore, we reviewed data on the individual projects included in the agencies' total EHS research funding in 2010. To assess the reliability of the agencies' data, we obtained information regarding the data and the information systems used to produce and store them. We also reviewed related supporting documentation and consulted with agency officials. From our assessment, we determined that the data were sufficiently reliable for our purposes. We analyzed these data and consulted with agency officials to assess whether the agencies had appropriately categorized as EHS research the projects they reported as program component area 7—research "primarily directed at understanding the EHS impacts of nanotechnology development and corresponding risk assessment, risk management, and methods for risk mitigation." We also reviewed our 2008 report in this area,[11] which found that 22 of 119 fiscal year 2006 projects reported as EHS research were miscategorized, and recommended that the Director of OSTP, in consultation with the Directors of NNCO and OMB, provide better guidance to agencies on how to report nanotechnology EHS research. To identify the nanomaterials studied by the fiscal year 2010 EHS research projects, we reviewed qualitative project data such as abstracts for the projects we determined were primarily directed at EHS and consulted with agency officials. We then grouped the nanomaterials into five categories.[12]

To determine the extent to which the NNI agencies collaborate with stakeholders on nanotechnology EHS research and related strategies, we (1) discussed with agency officials how their agencies collaborate on nanotechnology EHS research and NNI's role in facilitating that collaboration and (2) obtained documentation on these collaborative efforts. We conducted a review of formal collaborative efforts that focused on nanotechnology EHS research initiated from February 2008 to October 2011. We administered a web-based questionnaire to a nonprobability sample of 223 nonfederal stakeholders, including those affiliated with academia, companies, nongovernmental organizations (NGO), and state and local governments, to obtain their views on collaboration with the NNI member agencies on EHS research and the NNI EHS research strategies. Of the 223, 138 completed the questionnaire, for an overall response rate of 62 percent. This sample included individuals who had expertise in the field of nanotechnology, had interacted with the NNI in the past few years, or who were representatives of organizations and companies suggested to us through our scoping interviews.

To determine the extent to which the NNI strategy documents address desirable characteristics of national strategies, we analyzed three key NNI strategy documents related to nanotechnology EHS for the presence of six desirable characteristics we previously identified for national strategies. We also reviewed NNI member agencies' management documents for the use of performance information related to nanotechnology EHS research and the extent to which they were linked with the NNI national priorities. Appendix I presents a more detailed description of our scope and methodology.

We conducted this performance audit from February 2011 through May 2012 in accordance with generally accepted government auditing standards. Those standards require that we plan and perform the audit to obtain sufficient, appropriate evidence to provide a reasonable basis for our findings and conclusions based on our audit objectives. We believe that the evidence obtained provides a reasonable basis for our findings and conclusions based on our audit objectives.

BACKGROUND

Nanomaterials come in a variety of forms based both on their chemical composition and physical structure. For example, carbon-based nanomaterials can be produced in a number of physical structures such as sheets (graphene), tubes (carbon nanotubes), and particles (carbon black). Nanomaterials can enter the marketplace as materials themselves, as intermediates that either have nanoscale features or incorporate nanomaterials, and as final nano-enabled products.

The use of nanomaterials in commercial applications raises questions about the potential risks that might arise from exposures to nanomaterials and the differences in exposure during their manufacture, use, and disposal. For example, some, but not all, research studies have shown adverse respiratory or cellular effects in animals exposed to some types of carbon nanotubes. Observed effects include early onset and persistence of pulmonary fibrosis and interference with cell division.[13] The risk posed by a material is a combination of the hazard or negative effect that material may have on an organism and the extent of the organism's exposure to that material. Therefore, a highly poisonous material—that is, one with high hazard—may nonetheless pose little risk if susceptible groups have little or no contact with the material. For instance, many household chemicals are hazardous to human health but pose little risk when exposure is limited by safe handling. Conversely, a material with relatively mild health effects may pose a large risk if people or the environment are exposed to large amounts or over prolonged periods.

Efforts to examine EHS risks are complicated by the variety of nanomaterials available for research and commerce. According to the National Research Council (NRC) of the National Academies, a nonprofit institution providing scientific advice, there are currently gaps in understanding what factors or underlying phenomena contribute to potential hazards from nanomaterials.[14] In 2010, we reported that size alone is not a sufficient indicator of the potential risk of a material.[15] Research on carbon nanotubes, for example, suggests that length, purity, and other factors may be involved as well. Similarly, data on potential exposures of people and the environment are incomplete. In the absence of clear indicators of risk, some organizations sponsoring research on nanomaterials have identified specific materials or classes of materials as high-priority targets for EHS testing based on availability and commercial interest. For example, as part of a testing program of the EHS risks of nanomaterials, the Organisation for Economic Co-operation and Development has developed a list of 13 representative manufactured nanomaterials now or soon to be in commerce as priority testing targets.[16,17]

Regulatory agencies, like FDA, CPSC, and EPA, may not have complete information on the uses or risks of some nanomaterials. In 2010, we reported that uncertainties persist about how to evaluate the safety of engineered nanomaterials in food and that nanomaterials may enter the food supply in certain products generally recognized as safe without FDA's knowledge.[18] We also reported in 2010 that EPA has taken a variety of actions to better understand and regulate the risks of nanomaterials but that the agency faces challenges that might impede its ability to regulate nanomaterials effectively.[19] For example, we reported that the Toxic Substances Control Act gives EPA authority to issue rules requiring companies to submit certain information about chemicals. EPA recently amended a regulation to require companies to report certain information regarding production of chemicals above certain thresholds but those thresholds may not capture nanomaterials if they are produced in amounts below the thresholds. EPA currently has no plans to further reduce the thresholds. In addition, according to officials at CPSC, which is charged with protecting the public from unreasonable risks of injury or death from thousands of types of consumer products, CPSC does not have the statutory authority to require pre-market approval for products, including those incorporating nanomaterials.

The NNI was codified in law by the 21st Century Nanotechnology Research and Development Act in 2003.[20] The act requires the President to implement a national nanotechnology program and charges the NSTC itself—

or through a subgroup—with overseeing the planning, management, and coordination of the program. The NSTC carries out these tasks through the NSET Subcommittee, which includes a co-chair from OSTP as well as representatives from the member agencies of the NNI. The NSET oversees working groups, including the Nanotechnology Environmental and Health Implications (NEHI) working group which supports federal activities to protect public health and the environment. Figure 1 provides an overview of the organizational structure of the NNI.

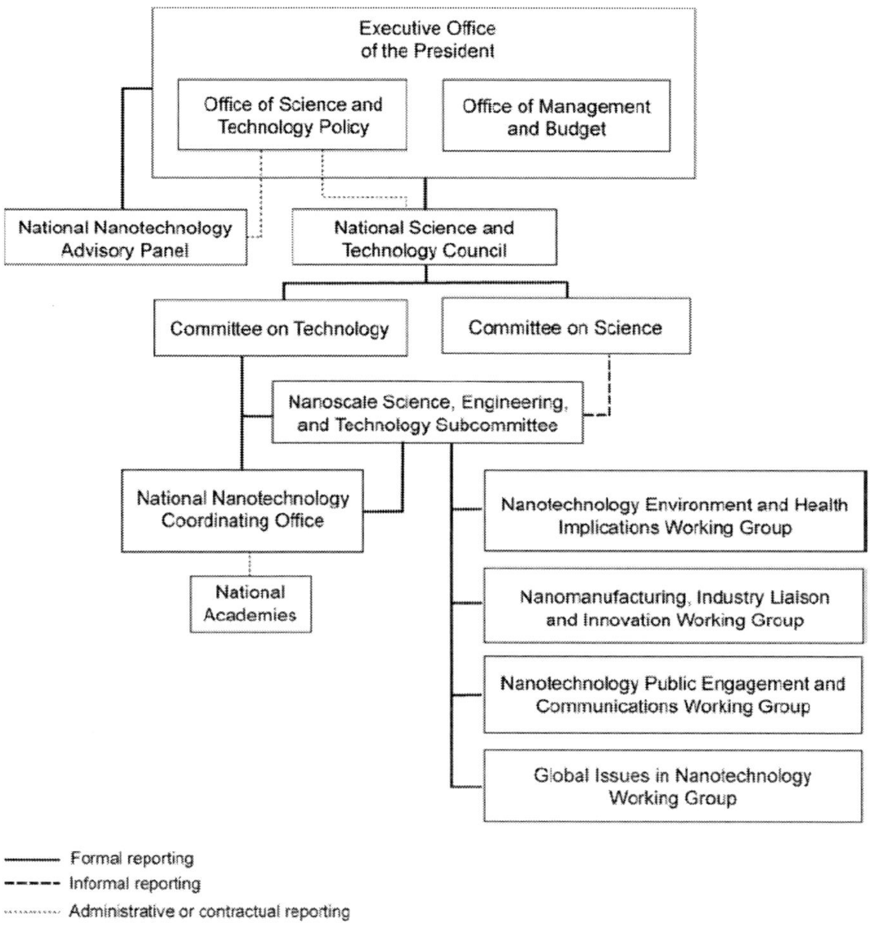

Source: NNI.

Figure 1. The NNI Structure.

The NNI does not fund research directly; rather, each of its member agencies determines its nanotechnology activities based on its individual mission and priorities. The NNI provides a framework for a comprehensive nanotechnology research and development program by establishing shared goals, priorities, and strategies among member agencies; and providing avenues for member agencies to leverage the resources of all participating agencies. The four goals of the NNI are to (1) advance a world-class nanotechnology research and development program; (2) foster the transfer of new technologies into products for commercial and public benefit; (3) develop and sustain educational resources, a skilled workforce, and the supporting infrastructure and tools to advance nanotechnology; and (4) support responsible development of nanotechnology. Efforts of the NNI member agencies are reported through triennial strategic plans and annual budget supplements. The act directs the NSTC, itself or through an appropriate subgroup it designates or establishes, to develop and update every 3 years a strategic plan to guide the activities of the program. The NSTC published its most recent strategic plan in February 2011.[21] In addition to the statutorily required strategic plan, in 2008, the NSTC published a *Strategy for Nanotechnology-Related Environmental, Health, and Safety Research* and updated this document in 2011.[22] The EHS strategies published in 2008 and 2011 expand on the goal of the responsible development of nanotechnology by describing the state of science and research needed to ensure that nanotechnology provides maximum benefits to the environment and human well-being. The act also requires the NSTC to prepare an annual report to be submitted to congressional committees on the national nanotechnology program's budget and an analysis of the progress made toward achieving the goals and priorities established for the program, among other things. In the NNI Supplements to the President's Budget, the NSTC reports overall NNI spending and describes research efforts and investments within eight program component areas, as seen in table 1. These program component areas provide an organizational framework for categorizing the activities of the NNI. Data presented in the NNI annual supplements is collected from the NNI member agencies by OMB as part of the annual budget formulation process. The act requires triennial external reviews of the national nanotechnology program. Specifically, the act requires the NNCO to contract with the NRC to conduct the triennial evaluations of the national nanotechnology program. The NRC draws on expertise from outside government, including from academia, companies, and NGOs.

Table 1. Program Component Areas

No.	Title	Description
1	Fundamental Nanoscale Phenomena and Processes	Discovery and development of fundamental knowledge pertaining to new phenomena in the physical, biological, and engineering sciences that occur at the nanoscale. Elucidation of scientific and engineering principles related to nanoscale structures, processes, and mechanisms.
2	Nanomaterials	Research aimed at the discovery of novel nanoscale and nanostructured materials and at acomprehensive understanding of the properties of nanomaterials (ranging across length scales, and including interface interactions). Research and development (R&D) leading to the ability to design and synthesize, in a controlled manner, nanostructured materials with targeted properties.
3	Nanoscale Devices and Systems	R&D that applies the principles of nanoscale science and engineering to create novel, or to improve existing, devices and systems. Includes the incorporation of nanoscale or nanostructured materials to achieve improved performance or new functionality. To meet this definition, the enabling science and technology must be at the nanoscale, but the systems and devices themselves are not restricted to that size.
4	Instrumentation Research, Metrology, and Standards for Nanotechnology	R&D pertaining to the tools needed to advance nanotechnology research and commercialization, including next-generation instrumentation for characterization, measurement, synthesis, and design of materials, structures, devices, and systems. Also includes R&D and other activities related to development of standards, including standards for nomenclature, materials characterization and testing, and manufacture.
5	Nanomanufacturing	R&D aimed at enabling scaled-up, reliable, and cost-effective manufacturing of nanoscale materials, structures, devices, and systems. Includes R&D and integration of ultra-miniaturized top-down processes and increasingly complex bottom-up or self-assembly processes.
6	Major Research Facilities and Instrumentation Acquisition	Establishment of user facilities, acquisition of major instrumentation, and other activities that develop, support, or enhance the nation's scientific infrastructure for the conduct of nanoscale science, engineering, and technology R&D. Includes ongoing operation of user facilities and networks.
7	Environment, Health, and Safety	Research primarily directed at understanding the environmental, health, and safety impactsof nanotechnology development and corresponding risk assessment, risk management, and methods for risk mitigation.
8	Education and Societal Dimensions	Education-related activities such as development of materials for schools, undergraduate programs, technical training, and public communication, including outreach and engagement. Research directed at identifying and quantifying the broad implications of nanotechnology for society, including social, economic, workforce, educational, ethical, andlegal implications.

Source: NSTC, 2011 NNI Strategic Plan.

The NRC completed reviews in 2002, 2006, and 2009 based on the work of 15 to 23 panelists chosen by the NRC and identified in the review. In addition, the NRC also published an independent research strategy to address EHS aspects of nanomaterials in 2012.[23]

The act also requires the President to establish or designate a National Nanotechnology Advisory Panel. The advisory panel, by statute, must consist primarily of members from academic institutions and industry, and panel members must be qualified to provide advice and information on nanotechnology research, development, demonstrations, education, technology transfer, commercial application, or societal and ethical concerns. Since 2004, the President has designated the President's Council of Advisors on Science and Technology (PCAST) to function as the advisory panel.

PCAST members are generally senior managers in major corporations and academia selected for diverse expertise in science and technology issues. The advisory panel is required to report not less frequently than once every 2 fiscal years on its assessment of the national nanotechnology program and recommendations for ways to improve the program.

The advisory panel has produced three assessments to date, in 2005, 2008, and 2010. The first and second assessments were authored by a subset of PCAST membership, and the third assessment was authored by a working group of three PCAST members and additional external experts, but PCAST as a whole approved the assessments.

The first and second assessments created Nanotechnology Technical Advisory Groups, which were comprised of approximately 40 members who provided written responses to questionnaires developed by PCAST. The membership of these groups is not identified in the assessments. The external members of the third assessment's working group were selected by PCAST members and are identified in PCAST's assessment.[24]

Funding of EHS Research by NNI Member Agencies Has Increased, but Quality of Funding Data Is Uncertain

NSTC reported more than a doubling of funding for nanotechnology EHS research by NNI member agencies from fiscal years 2006 to 2010.[25] Reported EHS funding also rose as a percentage of total NNI funding during this period, ending up at about 5 percent in 2010. We also identified several reporting problems related to the continued absence of detailed guidance on how agencies should report funding for their nanotechnology research, raising concerns about the quality of EHS funding data reported.

EHS Research Funding by NNI Member Agencies More Than Doubled From 2006 to 2010 and Amounted to About 5 Percent of Total NNI Funding in 2010

From fiscal years 2006 to 2010, NSTC reported more than a doubling of NNI member agencies' funding for nanotechnology EHS research in the NNI Supplements to the President's Budget—from approximately \$38 million[26] to \$90 million.[27] As shown in figure 2, most of this funding was reported to be by NSF, NIH, and EPA, and the largest increases in reported EHS research funding over this period were at NIH and EPA. Figure 3 shows the percentage of NNI funding represented by each program component area (PCA) for all NNI member agencies funding nanotechnology research. Most of the funding reported was for PCA 1 (fundamental nanoscale phenomena and processes), PCA 2 (nanomaterials), and PCA 3 (nanoscale devices and systems). Reported EHS funding rose as a percentage of total NNI funding from fiscal years 2006 to 2010, ending up at about 5 percent in 2010. Specifically, for fiscal year 2010, funding for EHS research was \$90 million, while total 2010 nanotechnology funding was \$1.9 billion, as reported in the NNI Supplement to the President's 2012 Budget.

Reporting Problems Raise Concerns about Quality of EHS Research Funding Data

Based on our review of the NNI member agencies' EHS research funding data published by NSTC in the NNI Supplements to the President's Budget and detailed project data we obtained from the seven selected agencies for fiscal year 2010, we identified several reporting problems that raise concerns about the accuracy, consistency, and completeness of funding data reported and make it difficult to assess changes in federal funding for nanotechnology EHS research over time. We found that (1) for 18 percent of 2010 projects reported as EHS research, it was not clear that the projects were primarily directed at EHS risks; (2) agencies vary in how they report funding for portions of projects that address multiple program component areas or include non-nanotechnology research; (3) agencies did not always report comparable budget data; and (4) NNI does not report updated data after publication of the NNI budget supplements, and the reported data may not reflect all EHS funding.

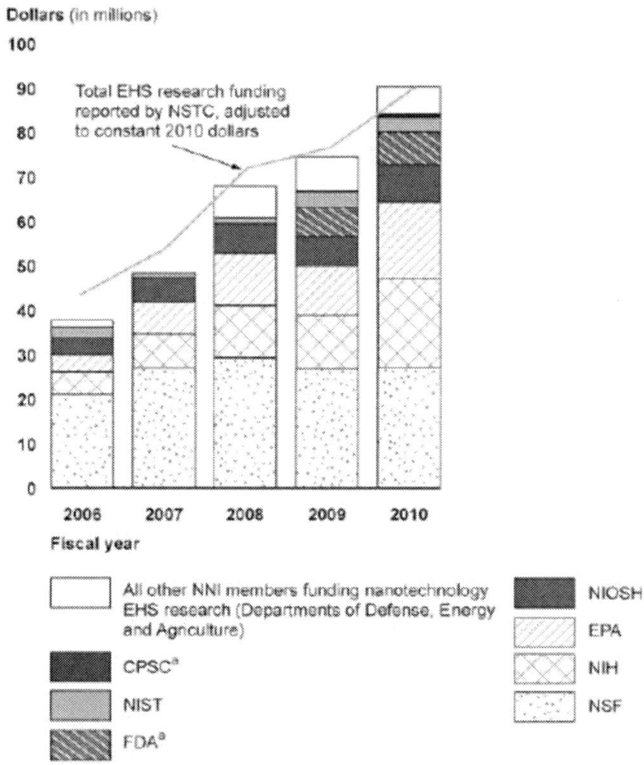

Note: This figure does not include research NSTC separately reported as funded by the American Recovery and Reinvestment Act of 2009 (Recovery Act) (Pub. L. No. 111-5 (2009)). The Recovery Act was enacted to, among other things, preserve and create jobs and promote economic recovery. For fiscal year 2009, NSTC reported additional funding for EHS research funded by the Recovery Act, as follows: NSF, $3.4 million; NIH, $8.4 million; and Department of Energy, $0.2 million. For fiscal year 2010, NSTC did not separately report nanotechnology funding from Recovery Act appropriations.

[a]The NNI Supplements to the President's Budget did not include nanotechnology funding data for CPSC and FDA prior to fiscal year 2009.

Figure 2. Nanotechnology EHS Research Funding Reported under Program Component Area 7 by NSTC for All NNI Member Agencies, Fiscal Years 2006 through 2010.

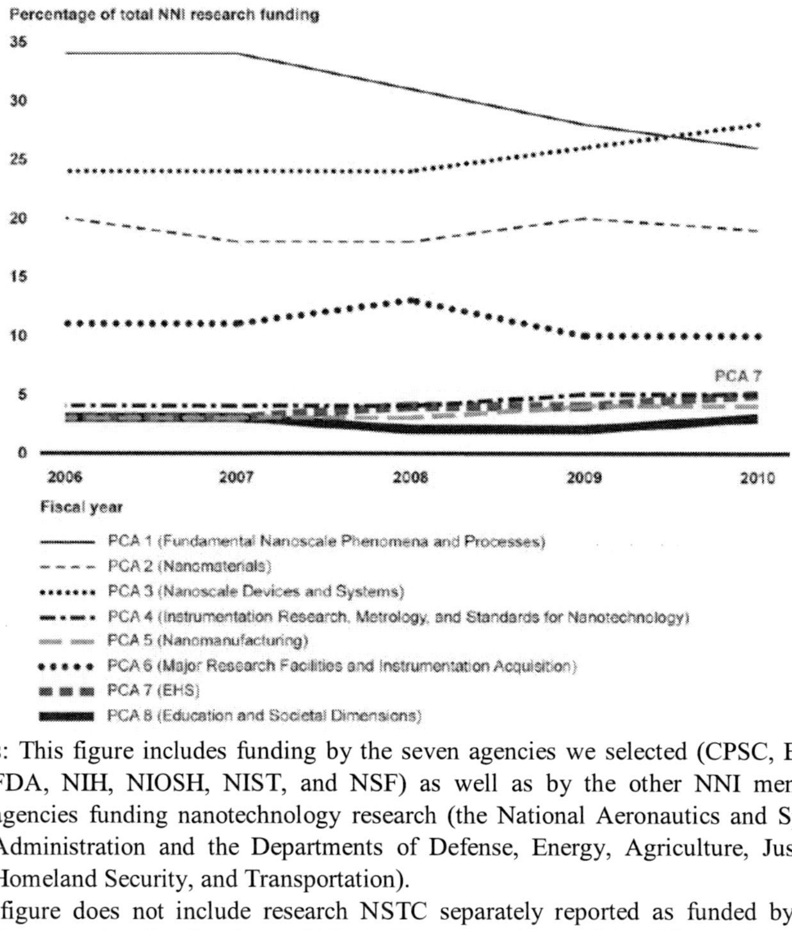

Notes: This figure includes funding by the seven agencies we selected (CPSC, EPA, FDA, NIH, NIOSH, NIST, and NSF) as well as by the other NNI member agencies funding nanotechnology research (the National Aeronautics and Space Administration and the Departments of Defense, Energy, Agriculture, Justice, Homeland Security, and Transportation).

This figure does not include research NSTC separately reported as funded by the Recovery Act. For fiscal year 2009, NSTC reported an additional $511 million for activities funded by the Recovery Act, as follows: PCA 1, $131 million; PCA 2, $178 million; PCA 3, $68 million; PCA 4, $12 million; PCA 5, $29 million; PCA 6, $73 million; PCA 7, $12 million; and PCA 8, $9 million. For fiscal year 2010, NSTC did not separately report agencies' funding from the Recovery Act appropriations.

Figure 3. Percentage of Total NNI Research Funding Represented by Each Program Component Area, Fiscal Years 2006 through 2010.

Eighteen Percent of Projects Reported as EHS Research Were Not Clearly Directed at EHS Risks

Of the 236 projects that the seven agencies reported to us as EHS research for fiscal year 2010, we determined that, for 43 projects (18 percent), it was not clear that the projects met the definition for PCA 7— research primarily directed at the EHS impacts of nanotechnology development and corresponding risk assessment, risk management, and methods for risk mitigation. As shown in table 2, these projects, 20 of which were funded by NSF, accounted for approximately $15 million of the reported EHS funding. NSF officials told us that the agency reported projects as EHS research if they had some relevance to EHS instead of reporting only those projects primarily directed at EHS.

We also found that, for some projects reported by other agencies, studying EHS risks was not the primary purpose of the projects. Instead, some projects focused on, for example, exploring how nanotechnology could be used in various applications, such as remediation of environmental contamination, detection of chemical hazards or pathogens, or medical imaging. In addition, some other projects included some research relevant to understanding EHS impacts of nanotechnology, but it was not clear that the study of EHS issues was the primary purpose. For example, some projects studied various nanotechnology-enabled drugs, but it was not clear to what extent the research was directed at the safety of the drugs versus their efficacy. Other projects were directed at instrumentation and nanotechnology measurement issues that had some relevance for EHS research, but it was not clear whether these should be reported as PCA 7 (EHS research) or PCA 4 (instrumentation research, metrology, and standards for nanotechnology).

As previously noted, we recommended in our 2008 report that the Director of OSTP, in consultation with the Directors of NNCO and OMB, provide better guidance to agencies on how to report nanotechnology EHS research.[28] OSTP generally concurred with the report's findings and agreed to review the manner in which agencies respond to the current guidance. However, we found that as of February 2012, updated guidance had not been issued, and the definition of each program component area as documented in OMB Circular A-11 remained the only written guidance available to the agencies for reporting their nanotechnology EHS research.[29] OMB Circular A-11 does not address what types of projects should or should not be categorized as EHS research, such as how to determine if a project is "primarily" directed at EHS impacts or how to report projects focused on environmental or health-related applications of nanotechnology. For example, with respect to drugs, HHS commented that the Federal Food, Drug, and Cosmetic Act requires that both safety and effectiveness be evaluated, in relation to one another. As HHS noted, both traits are relevant to the overall health impact of a drug product.

Table 2. GAO Analysis of Nanotechnology Projects Reported by Selected Agencies as EHS Research, Fiscal Year 2010

	Projects seven agencies to GAO as EHS	reported research		GAO analysis of agencies' projects		
			Primarily directed at EHS		Not clear that projects were primarily directed at EHS	
Agency	Number[a]	Funding (dollars in millions)	Number	Funding (dollars in millions)	Number	Funding (dollars in millions)
CPSC	8	$0.5	8	$0.5	0	$0
EPA[b]	32	17.6[c]	29	16.9	3	0.8
FDA	29[d]	5.3[c]	20	3.8	9	1.5
NIH	58	27.7[e]	55	26.4[f]	3	1.3[g]
NIOSH	49	9.8[h]	47	9.7	2	0.1
NIST	11	3.4	5	1.6	6	1.8
NSF	49	27.1	29	17.1	20	10.0
Total[i]	236	$91.4[j]	193	$76.0	43	$15.5

Source: GAO analysis of obligations data reported to us by seven agencies.

Notes: The data shown in this table may differ from that reported by NSTC in the NNI Supplement to the President's 2012 Budget, as described below.

[a] For some projects at NIH, NIOSH, and NIST, we grouped together some of the agencies' data entries and counted them together for the purposes of our analyses. Consequently, the numbers of projects reported here may not match the number reported by these agencies elsewhere. Appendix I presents more details on how we grouped projects at these agencies.

[b] According to an EPA official, the total EHS research funding reported to us for fiscal year 2010 represents actual obligations, but the amounts reported for individual projects are estimates.

[c] The total for this agency differs from that reported by NSTC in the NNI Supplement to the President's 2012 Budget because the agency reported its budget authority to OMB for the budget supplement and reported obligations to us.

[d] FDA reported a total of 30 EHS research projects to us; however, for one of those projects, FDA reported that no funds were obligated in fiscal year 2010. Therefore, we excluded that project from our analysis.

[e] The total for this agency differs from that reported by NSTC in the NNI Supplement to the President's 2012 Budget because NIH included in the data reported to us its fiscal year 2010 obligations of funds appropriated by the Recovery Act, which it did not report to OMB for the supplement.

[f] We added $35,044 to NIH's reported total because we identified additional funding for a subproject that we determined should have been included. NIH officials told us that the sub-project was not included because the project abstract and other associated text did not contain any of the key words for the computerized search of projects that NIH uses to develop its list of EHS research.

[g] For two additional NIH projects, we found that the amount of funding for research primarily directed at EHS was overstated. For each of these projects, NIH included the funding for one or more subprojects that were not focused on nanotechnology. We included the overstated amounts in the dollar value for projects where it was not clear that the research was primarily directed at EHS.

[h] The total for this agency differs from that reported by NSTC in the NNI Supplement to the President's 2012 Budget because NIOSH updated its data after initially reporting it to OMB.

[i] Totals may not add because of rounding.

[j] The total dollar value of $91.4 million reported to us differs from the total of $90.2 million reported by NSTC in the NNI Supplement to the President's 2012 Budget for two reasons. First, this table includes only the seven agencies we reviewed, not all agencies that funded EHS research in 2010. Second, four of the seven agencies we reviewed reported data to us that differed from those reported by NSTC.

However, OMB Circular A-11 does not address whether funding for research on the effectiveness of nanotechnologyenabled drugs should be considered nanotechnology EHS research under the definition of PCA 7. It also does not address how to report EHS-related projects that are focused on instrumentation and metrology, and we found that agencies have used different criteria to determine the program component area under which to report such projects as well.

NNCO and FDA officials told us that the reporting of EHS research was a regular agenda item at meetings of the NEHI working group during development of the 2011 NNI EHS research strategy. However, the strategy used information reported via a different OMB data call, one that NNCO and OMB officials said was much more detailed and cast a broader net than the annual nanotechnology data calls OMB conducts for the annual NNI budget supplements (the "crosscut"). The 21st Century Nanotechnology Research and Development Act requires annual reporting of spending for each program component area, and the crosscut collects EHS research funding data based on the definition of PCA 7— research primarily directed at the EHS impacts of nanotechnology development and corresponding risk assessment, risk management, and methods for risk mitigation. However, the data call for the 2011 research strategy was, according to OMB's instructions to agencies, "not limited to research and development efforts whose primary purpose is to understand and address potential risks to health and to the environment posed by this technology—as is called for in the overall crosscut for nanotechnology." Consequently, discussions of how to report projects for use in developing the 2011 research strategy do not necessarily provide guidance to agencies on how to determine whether their research meets the definition of PCA 7, or how to apportion funding for projects that address multiple program component areas.

Agencies Vary in How They Report Funding for Portions of Projects

In our 2008 report, we found that neither NSET nor OMB had provided guidance on whether or how to apportion funding for a single research project to more than one program component area, if appropriate.[30] In this review, we found that agencies have different approaches to reporting funding for projects that address multiple program component areas or include non-nanotechnology research, and the guidance available to the agencies—the PCA definitions in OMB Circular A-11—still does not address this. We found that, for fiscal year 2010 projects where only a portion of the project was focused on nanotechnology EHS research, NIST and NSF assigned only a portion of the total funding to PCA 7. At NIH, grants or contracts may be either single projects or multiproject "parent" grants or contracts, which have multiple individual subprojects. According to NIH officials, NIH policy does not allow the funding for a single project or subproject to be subdivided across program component areas, and NNI policy requires that the individual totals of the program component areas sum to the agency's total nanotechnology funding. As a consequence of these two policies, if a grant or contract is a single project, then NIH reports the entire value of the individual grant or contract in only one program component area. NIH officials told us that if a grant or contract has subprojects, NIH reports the funding for an individual subproject in only one program component area, but the funding for the different subprojects can be reported in different program component areas, as applicable.

At NIOSH, agency officials told us that, for projects that include non-nanotechnology research, they track and report the nanotechnology research portions separately and that the agency reports all of its nanotechnology research as EHS because NIOSH's mission aligns with PCA 7. The NIOSH mission includes conducting research on the potential implications of nanotechnology on worker health and safety, as well as conducting research on potential applications of nanotechnology to solve worker health and safety problems. FDA officials also told us that they report all of their regulatory science research regarding nanotechnology as EHS because it is conducted to support the agency's mission, which they said aligns with PCA 7. As described in FDA's *Strategic Priorities* document, the agency's mission includes ensuring both the safety and the effectiveness of drugs, biological products, and medical devices, as well as the safety of foods and cosmetics.[31]

Agencies Did Not Always Report Comparable Budget Data

Data on nanotechnology funding published in the annual NNI Supplements to the President's Budget are not comparable over time or across agencies because the data do not always represent the same type of budget information. The NNI budget supplements report funding as "actual agency investments" but do not specify what type of data are reported as actual agency investments, and until recently, OMB's instructions to agencies for providing these data did not specify what should be reported.[32] Officials from OMB told us that agencies report budget authority for their actual agency investments,[33] but officials from the seven agencies we reviewed generally told us that they had reported obligations as their actual agency investments for fiscal years 2006 to 2010.[34] However, an agency's budget authority and obligations are not always the same amounts in a given year. Furthermore, agencies may not always report the same type of budget information from year to year. For example, for fiscal year 2009, FDA reported obligations, but then reported budget authority for fiscal year 2010 because obligations data were not available, according to FDA officials.

NNI Does Not Report Updated Data after Publication of Budget Supplements, and Reported Data May Not Reflect All EHS Funding

Funding data published by NSTC in the NNI Supplements to the President's Budget are not always final because agencies sometimes make changes to their data after they are published as actual agency investments in the budget supplements. Consequently, outdated data can remain available in NNI documents and on the NNI website, which publishes nanotechnology research funding data from the budget supplements in response to OMB's 2009 Open Government Directive. 35 For example, NIOSH officials told us that their fiscal year 2010 actual agency investment in EHS research reported in the NNI Supplement to the President's 2012 Budget was an estimate because the timing of OMB's call for the data was earlier than usual. According to NIOSH officials, a second data call occurred later in the fiscal year, at which time the agency was able to report an accurate actual funding level. However, as of February 2012, NIOSH's estimated data have not been updated on the NNI website. In addition, we found that when NIST reported its 2009 nanotechnology research funding from the American Recovery and Reinvestment Act of 2009 (Recovery Act) ($43.4 million), this was an estimate, and the funding was actually obligated over 2 years ($6 million obligated in fiscal year 2009, and $37.4 million obligated in fiscal year 2010). However, as of February 2012, NIST's estimated data have not been updated on the NNI website.

In some instances, the reported data may not reflect all EHS research funding. We found that the fiscal year 2010 data reported by NSTC in the NNI Supplement to the President's 2012 Budget did not include all 2010 nanotechnology funding from Recovery Act appropriations, and OMB did not request that information from agencies in its call for 2010 data.[36] We found that NIH,[37] NIST,[38] and NSF[39] together obligated a total of $131.2 million for nanotechnology research in 2010 with Recovery Act appropriations (including $8 million for EHS research by NIH).[40] However, while NSF included its Recovery Act funds in the funding reported in the NNI Supplement to the President's 2012 Budget, NIH and NIST did not. In addition, some EHS research funding may not be included in the budget supplements because of agency-specific reporting methods.[41]

EHS Research Funded in 2010 Focused on Metal- and Carbon-Based Nanomaterials

In fiscal year 2010, EHS research at the seven NNI member agencies we reviewed focused on two categories of nanomaterials more than others—metal- and carbon-based nanomaterials, which are used in a variety of applications, including electronics, consumer products such as sunscreens, medical products, and protective coatings (see fig. 4). NNI has not prioritized nanomaterials for EHS research, but in October 2011 it outlined criteria for its member agencies to use in doing so. It is not yet clear what effect these criteria will have on how agencies prioritize the nanomaterials they focus on in their EHS research.

2010 Research Focused on Carbon Nanotubes, Silver, and Titanium Dioxide

As shown in figure 5, we found that in the 193 fiscal year 2010 research projects we determined were primarily directed at EHS, the nanomaterials that were most frequently the focus of this research were carbon nanotubes (70 projects), nanosilver (60 projects), and nanoscale titanium dioxide (48 projects).[42]

- Carbon nanotubes are nanoscale cylinders consisting of seamlessly "rolled" sheets of graphene, a form of carbon. They are extraordinarily strong, flexible, lightweight, heat resistant, and have high electrical conductivity. Carbon nanotubes are currently used in a variety of applications including conductive coatings for touchscreens and solar cells, and sporting goods such as bicycle frames and baseball bats. According to a 2009 NIOSH report, carbon nanotubes have been shown to produce adverse effects in the respiratory systems of rats.[43]

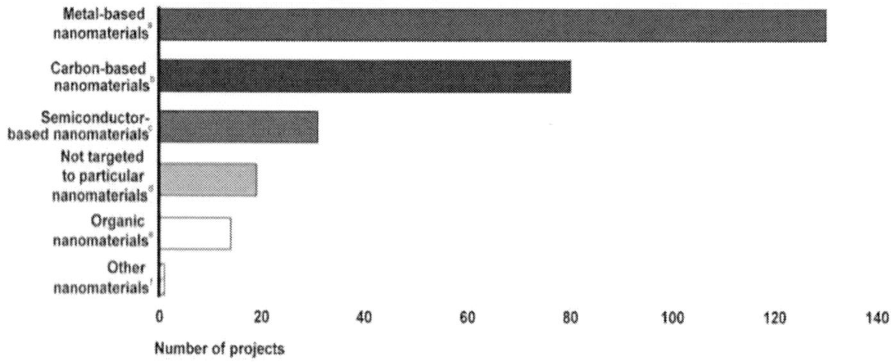

Source: GAO analysis of selected agencies' data on EHS research funded in 2010.

Notes: The figure is based upon our analysis of the 193 projects we determined were primarily directed at EHS. It does not include the 43 projects for which it was not clear to us that the research was primarily directed at EHS. The numbers of projects shown do not total 193 because projects often targeted more than one category of nanomaterials.

[a] Metal-based nanomaterials consisted of silver, gold, iron-based nanomaterials, various metal oxides (titanium, zinc, cerium, and aluminum oxides), and other metal-based nanomaterials.

[b] Carbon-based nanomaterials consisted of carbon nanotubes, fullerenes, and other carbon-based nanomaterials. Spherical and ellipsoidal carbon nanomaterials are referred to as fullerenes, while cylindrical ones are called nanotubes.

[c] Semiconductor-based nanomaterials consisted of quantum dots, silicon-based nanomaterials, and other semiconductor-based nanomaterials. Quantum dots are closely packed semiconductor crystals composed of hundreds or thousands of atoms.

[d] Examples of projects not targeted to particular nanomaterials are projects that funded large research centers on the environmental implications of nanotechnology, or projects that purchased data on the market for nanomaterials in order to guide exposure assessment studies by providing information on potential consumer exposures.

[e] Organic nanomaterials consisted of dendrimers, other polymers, and other organic-based nanomaterials. Dendrimers are branched nanosized polymers. The surface of a dendrimer has numerous chain ends, which can be tailored to perform specific chemical functions.

[f] Other nanomaterials consisted solely of nanoclays (one project).

Figure 4. Categories of Nanomaterials Studied by Seven NNI Member Agencies' EHS Research Projects, Fiscal Year 2010.

- Nanosilver is a metal-based nanomaterial that, according to a 2010 EPA report, is currently used in an increasing number of consumer and medical products because of its remarkably strong antimicrobial properties.[44] For example, EPA reported that nanosilver is being incorporated into clothing, food contact materials such as packaging and cutting boards, household appliances, cosmetics, and children's toys. It is also used in industrial processes because of its ability to catalyze many reactions. According to EPA's 2010 report, there is clear evidence that nanosilver is toxic to aquatic and terrestrial organisms and may be detrimental to human health. In this report, EPA noted that several studies have shown that nanosilver can be released into the wastewater stream during washing, such as from socks containing nanosilver. The nanosilver released may disrupt the helpful bacteria used in wastewater treatment processes or be released into the environment.
- Nanoscale titanium dioxide is a metal-based nanomaterial used in sunscreens, protective coatings, and other materials to manage heat and light by blocking UV light from the sun's rays. It is also being added to paints, cements, windows, tiles, and other products for its sterilizing and deodorizing properties and is being used for antifogging coatings and self-cleaning windows. In addition, it is being investigated for use in removing contaminants from drinking water. According to a 2011 NIOSH report, nanoscale titanium dioxide is a potential occupational carcinogen when inhaled.[45] Regarding exposure to this nanomaterial in sunscreens, a 2010 FDA publication found that nanoscale titanium dioxide included in a formulation similar to currently marketed sunscreens is unlikely to significantly penetrate human skin.[46]

Effect of Recent NNI Criteria for Agencies' Prioritization of Nanomaterials Is Not Yet Clear

NNI has not prioritized nanomaterials for EHS research, but the 2011 NNI EHS research strategy outlines criteria to assist its member agencies in doing so. The main criteria NNI proposed are the extent to which (1) study of a particular nanomaterial may provide a major contribution to the research knowledge base, and (2) nanomaterials and nanotechnologyenabled products may pose a safety concern to workers, consumers, and the environment.

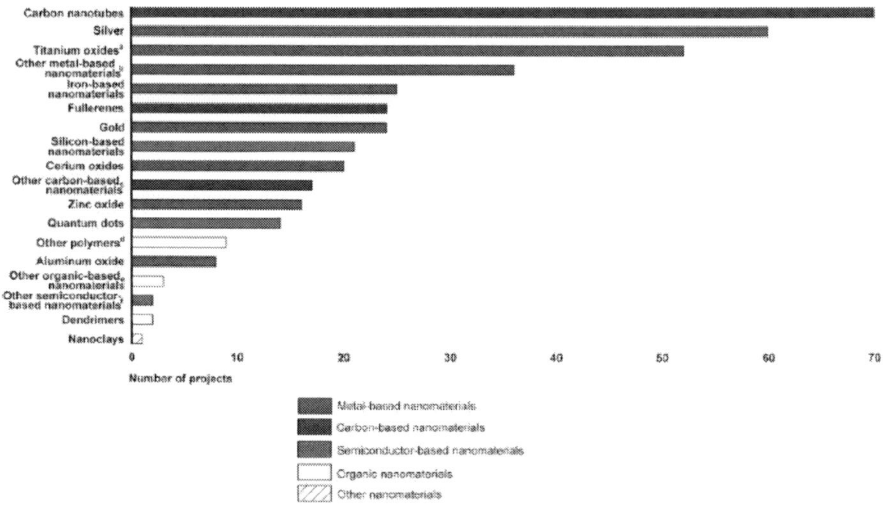

Source: GAO analysis of selected agencies' data on EHS research funded in 2010.

Notes: The figure is based upon our analysis of the 193 projects we determined were primarily directed at EHS. It does not include the 43 projects for which it was not clear to us that the research was primarily directed at EHS, nor those not targeted to particular nanomaterials. The numbers of projects shown do not total 193 because projects often targeted more than one nanomaterial.

[a] Of the 52 projects that targeted titanium oxides, at least 48 projects studied titanium dioxide. Two projects studied titanium oxides, but it was not clear if these included titanium dioxide.

[b] Other metal-based nanomaterials consisted of various metals, such as cobalt, copper, magnesium, manganese, nickel, and others, as well as some composites of multiple metals.

[c] Other carbon-based nanomaterials consisted of carbon nanofibers, carbon nanowires, carbon nanohorns, carbon black, diamond, graphene, nanoscale graphene platelets, plastic composite material containing carbon nanofibers, and soot nanoparticle agglomerates.

[d] Other polymers consisted of polystyrene, polyhedral oligomeric silsesquioxanes, nylon 6 nanofibers, plastics, polymeric nanovesicles, and poly(propargyl glycolid) nanoparticles.

[e] Other organic-based nanomaterials consisted of flourescein, sucrose, and dioctyl phthalate particles.

[f] Other semiconductor-based nanomaterials consisted of selenium nanoparticles and cadmium and lead selenides/sulfides.

Figure 5. Nanomaterials Studied by Seven NNI Member Agencies' EHS Research Projects, Fiscal Year 2010.

NNI further identified five criteria for assessing whether a material may pose a safety concern: (1) the nanomaterial's potential for hazard, (2) likelihood of exposure, (3) high reactivity, (4) biological novelty, and (5) the involvement of the material in an event that produced health or environmental impacts. However, because the criteria were published in October 2011, it is too soon to tell how they will influence NNI member agencies' decisions about which nanomaterials to prioritize.

Furthermore, it is unclear if the information needed to use the NNI criteria is available. As we reported in May 2010, predicting and assessing the potential hazards, exposures, and resulting risks from nanomaterials is difficult, and current understanding of nanomaterial toxicity and exposure is limited. For example, the findings from completed toxicity studies of a nanomaterial constructed in one manner may not be applicable to understanding the risks posed by the same nanomaterial constructed in a different manner and, therefore, studies of similar nanomaterials may not be comparable.[47] Also, according to the NRC's 2012 draft strategy for nanotechnology EHS research, there is incomplete information on the effects of the array of nanomaterials used in products and a lack of information on effects of chronic exposures. NRC reported that most toxicity studies test a single material and usually focus on effects of acute exposures.[48]

In addition, reliable, comprehensive information is not readily available on the likelihood of exposure to nanomaterials. Consumers and workers are more likely to be exposed to nanomaterials that are already produced in large quantities or incorporated into a larger number of products, but information on the prevalence or production volumes of nanomaterials or nanotechnology-enabled products is currently limited. As previously noted in this report, FDA, CPSC, and EPA may not receive complete information on what nanomaterials are already in use in the case of certain products. CPSC officials told us that one way the agency collects information on the use of nanomaterials in products to better understand potential exposures is by purchasing information from commercial vendors. However, we found that obtaining such information from commercial vendors such as market research companies can be costly, and it is difficult to assess the reliability of such research because it can involve data and methods, including modeling based on various assumptions, that may be proprietary.

Another source of information on nanomaterials in commerce is the Woodrow Wilson International Center for Scholars' Project on Emerging Nanotechnologies, which maintains an inventory of nanotechnologybased products.[49] As of the March 2011 update to the list, the inventory contained approximately 1,300 nanotechnology-based products. However, the Wilson Center's list is not comprehensive, and it consists of consumer products.[50] As we reported in May 2010, occupational exposure is a particular concern because the exposure and risk to workers is potentially greater than the risk to consumers.[51] At present, though, there is little information on the exposure of workers to nanomaterials in the workplace.

NNI MEMBER AGENCIES HAVE COLLABORATED EXTENSIVELY ON NANOTECHNOLOGY EHS RESEARCH

The NNI member agencies have collaborated extensively with each other and nonfederal stakeholders on EHS research and strategies. The NNI member agencies participated in interagency efforts to develop joint strategies related to EHS research. They have also undertaken over 40 collaborative nanotechnology EHS research projects in recent years, signing agreements with both federal and nonfederal stakeholders. Many of the nonfederal stakeholders responding to our questionnaire rated their collaborative activities with NNI agencies as very or generally useful, although they have identified some challenges.

NNI Member Agencies Have Collaborated to Develop Joint Strategies

NNI member agencies have collaborated through the NSTC to develop joint EHS research strategies. The NSTC primarily coordinates nanotechnology EHS research through the NSET Subcommittee and the NEHI, an interagency working group. The NEHI's purpose includes providing for an exchange of information among agencies and with nonfederal stakeholders; facilitating the identification, prioritization, and implementation of nanotechnology EHS research; and managing the EHS interagency research strategy and facilitating its implementation. The 2011 NNI EHS research strategy was developed by NNI member agencies working through the NEHI working group. The strategy sets a common vision for nanotechnology EHS research, EHS research categories and needs, and key principles intended to assist the NNI member agencies make strategic decisions about research programs that will advance the NNI EHS research agenda while meeting their respective missions. According to several NNI member agency officials, individual agencies' implementation efforts of the 2011 NNI EHS research strategy are discussed at NEHI meetings. However, since the strategy was finalized in October 2011, it is too soon to tell to what extent it will be used to integrate the NNI member agencies' implementation plans. The 2011 NNI EHS research strategy updated and replaced the 2008 NNI *Strategy for Nanotechnology-Related EHS Research*, which had established common priority EHS research needs. The 2008 NNI *Strategy for Nanotechnology-Related EHS Research* was also developed by the NEHI working group and informed by earlier publications of NNI EHS research needs.

Officials from five of the NNI member agencies told us that they use the NEHI working group as a forum to collaborate on nanotechnology EHS research projects. For example, CPSC officials told us that they identify agencies with expertise on testing nanomaterials found in consumer products through discussions in the NSET Subcommittee and NEHI working group, which has led to collaborative testing of exposure and release of selected nanomaterials. This collaboration has resulted in data that support the respective missions of the collaborative agencies. Officials from the National Toxicology Program at NIH noted that they selected nanomaterials for study based in part on discussions in the NEHI working group and NNI sponsored workshops. Officials from NIST told us that NNI member agencies are also working through the NEHI working group to develop an inventory of collaborative EHS research activities. Collaborative activities related to EHS research have been reported in the 2011 NNI EHS research strategy and annual NNI Supplements to the President's Budget.

NNI Member Agencies Have Engaged in Numerous Collaborative EHS Research Projects

The NNI member agencies have recently initiated numerous nanotechnology EHS research projects in collaboration with other federal agencies and nonfederal stakeholders. We reviewed formal collaborative nanotechnology EHS research projects initiated from February 2008 to October 2011, while the 2008 NNI *Strategy for Nanotechnology-Related EHS Research* was in effect. The 2008 NNI *Strategy for NanotechnologyRelated EHS Research* called for the NEHI and the NNI member agencies to coordinate agency efforts to address priority research needs and, among others, identify opportunities for collaboration and joint development and use of resources where appropriate, facilitate partnerships with industry, and coordinate and support international efforts, support development of consensus-based standards, and facilitate wide dissemination of research results and other nonproprietary EHS information. During this period, NNI member agencies initiated 43 formal collaborative projects related to EHS research—24 interagency collaborations and 19 collaborations that included nonfederal stakeholders.

Most of the interagency collaborations were among the NNI member agencies included in our review, but a few were with other federal agencies such as the Department of Defense, the Department of Labor's Occupational Safety and Health Administration, and the Department of Agriculture. Nonfederal stakeholders included foreign governments, such as the United Kingdom (UK) and China; universities, such as the University of Massachusetts, Lowell and Rice University; and NGOs, such as the International Alliance for NanoEHS Harmonization (IANH) and the International Life Sciences Institute Research Foundation.[52]

NNI member agencies used these collaborative projects to extend their capability to achieve their individual missions. For example, CPSC collaborated with four agencies to conduct research on nanomaterials found in consumer products. NIOSH, an agency within the Centers for Disease Control and Prevention that conducts research and makes recommendations for the prevention of work-related injury and illness, partnered with entities that operate nanomanufacturing facilities to better understand the potential for occupational exposure to nanomaterials.

NIST, which aims to advance measurement science, standards, and technology, signed an agreement with the IANH to jointly create protocols for toxicological tests of selected representative nanoparticles, confirming inter-laboratory reproducibility and verifying the predictability of certain procedures.

Collaborative EHS research projects have resulted in transfers of funding as well as sharing of expertise, facilities, and other resources, as shown in the following examples:

- An interagency agreement between CPSC and NIOSH to study the pulmonary effects of titanium dioxide nanoparticles released from aerosol spray products involved CPSC providing the product to be tested and transferring funding to NIOSH to construct the testing equipment, with NIOSH providing expertise and staff time to run the tests and produce a report.

- An agreement among the NIH's National Cancer Institute, the National Institute of Environmental Health Sciences (NIEHS), and the National Institute of Biomedical Imaging and Bioengineering to develop, maintain, and operate a web-based nanomaterial registry specified that the National Institute of Biomedical Imaging and Bioengineering will develop the registry to provide consistent and curated information on the biological and environmental interactions of well-characterized nanomaterials, as well as links to associated publications, modeling tools, computational results, and manufacturing guidance from existing databases. The National Cancer Institute and the NIEHS provided funding. NIEHS also committed to provide input to the constitution of an advisory board and attend its meetings, as well as be represented in the nano registry project management team.
- Agencies have also issued joint solicitations for grant applications, such as one between EPA and a number of UK agencies to solicit proposals for research on environmental and health implications of nanotechnology. EPA and the UK agencies signed an agreement that notes (1) the funding of grants will be made consistent with the budget priorities of each party and (2) they are to work jointly to define the scientific priorities, issue the request, assist with peer review and selection of grantees, and disseminate the research results.

Details on these and other collaborative EHS research projects are provided in appendix II.

Nonfederal Stakeholders We Surveyed Benefited from NNI Collaboration but Identified Some Challenges

The 138 nonfederal stakeholders who responded to our questionnaire reported that they benefited from collaboration with NNI member agencies but faced some challenges. Most respondents indicated that the 2011 NNI EHS research strategy was very or somewhat effective at addressing the EHS research needs.

Respondents reported that they also obtained information on nanotechnology EHS risks from NNI member agencies and nongovernmental sources.

Many of the Respondents to Our Questionnaire Rated Collaboration with NNI Member Agencies on EHS Research as Generally or Very Useful

The following three formal mechanisms for collaboration with NNI member agencies were most frequently identified by respondents as generally or very useful to them: (1) joint data gathering and sharing; (2) joint research solicitations or funding of research consortia; and (3) competitive grants. See figure 6 for respondents' ratings of seven formal collaborative mechanisms we identified for them to rate. Some respondents who provided comments in response to optional open-ended questions also identified public-private partnerships, such as joint academic, government, and industry information exchange and research programs, as collaboration mechanisms the NNI member agencies should consider. A few respondents who commented in response to optional open-ended questions also cited benefits to informal collaboration with NNI agencies, such as discussions during workshops and conferences.

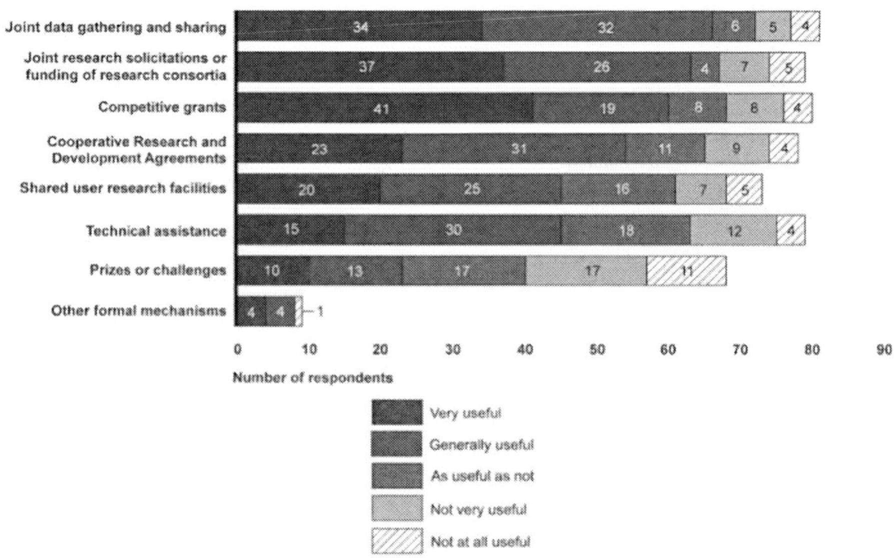

Source: GAO analysis of questionnaire of nonfederal stakeholders.

Note: Excludes respondents who selected "No basis to judge" and those who did not check a response.

Figure 6. Usefulness of Formal Collaboration Mechanisms for Nanotechnology EHS Research, According to Questionnaire Respondents.

Since both the 2011 NNI EHS research strategy and its predecessor, the 2008 NNI *Strategy for Nanotechnology-Related EHS Research*, called for partnerships with industry and other nonfederal stakeholders, we included questions about collaboration with NNI member agencies in our questionnaire. When asked about the usefulness of their collaboration with the NNI member agencies, more than half of respondents rated their collaborative EHS research or related activities for each of the NNI member agencies as generally or very useful, as seen in figure 7.

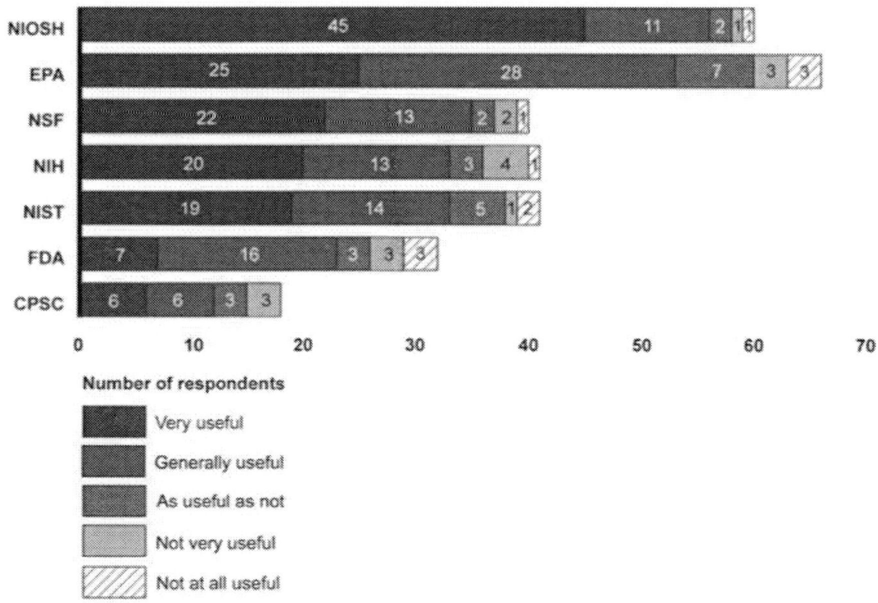

Source: GAO analysis of questionnaire of nonfederal stakeholders.

Note: Excludes respondents who selected "No basis to judge" and those who did not check a response.

Figure 7. Usefulness of Collaborative EHS Research or Related Activities with NNI Member Agencies, According to Questionnaire Respondents.

Respondents to Our Questionnaire Indicated a Number of Challenges to Collaboration with NNI Member Agencies Apply to Them

Stakeholders responding to our questionnaire indicated that certain challenges to collaboration with NNI member agencies we identified on EHS research apply to them. As seen in figure 8, more than two-thirds of those who rated the challenge of lack of funding and the challenge of limited awareness of collaboration opportunities indicated that each of these challenges apply to them. This is consistent with what we heard from representatives of a nanotechnology trade association and an NGO, who stated that there is limited funding for corporate EHS research and development efforts because they do not generate revenues. Some respondents who provided comments in response to optional open-ended questions stated that the private sector needs government funding for EHS research. A number of respondents also commented that government funding for EHS research was inadequate or an "afterthought" to other purposes. Some respondents also commented that funding seems to be targeted toward large, multiyear centers rather than smaller targeted projects that can help address near-term EHS needs.

More than half of those responding noted that each of the challenges of regulatory uncertainty, lack of standardization—such as different terminology or protocols—and concerns regarding disclosure of proprietary or confidential business information, apply to them. Some respondents cited regulatory uncertainty—that is, lack of a clear regulatory environment to enable commercialization and protect consumers and the general public—as a challenge to collaboration with their work with NNI member agencies on nanotechnology EHS research. For example, a number of respondents who provided comments in response to optional open-ended questions pointed to regulatory uncertainty as limiting the ability of companies to determine their EHS responsibilities. Some respondents also commented on difficulties in collaborating with NNI member agencies resulting from differences in expertise and regulatory approaches across different government agencies or even within the same agency. With regard to lack of standardization as a challenge to collaboration on nanotechnology EHS research, some respondents provided examples related to the lack of scientific consensus on the definition of nanomaterials, the testing methods for toxicology research, and common terminology across research and regulatory agencies.

Most Respondents to Our Questionnaire Rated the 2011 NNI EHS Research Strategy as Somewhat or Very Effective

When asked to rate the effectiveness of mechanisms we identified that NNI member agencies can use for obtaining input from nonfederal stakeholders on the development of the NNI EHS research strategies, respondents most frequently identified the following three mechanisms as being used somewhat effectively or effectively: (1) workshops, conferences, and other public speaking engagements; (2) advisory councils; and (3) the NNI web portal.

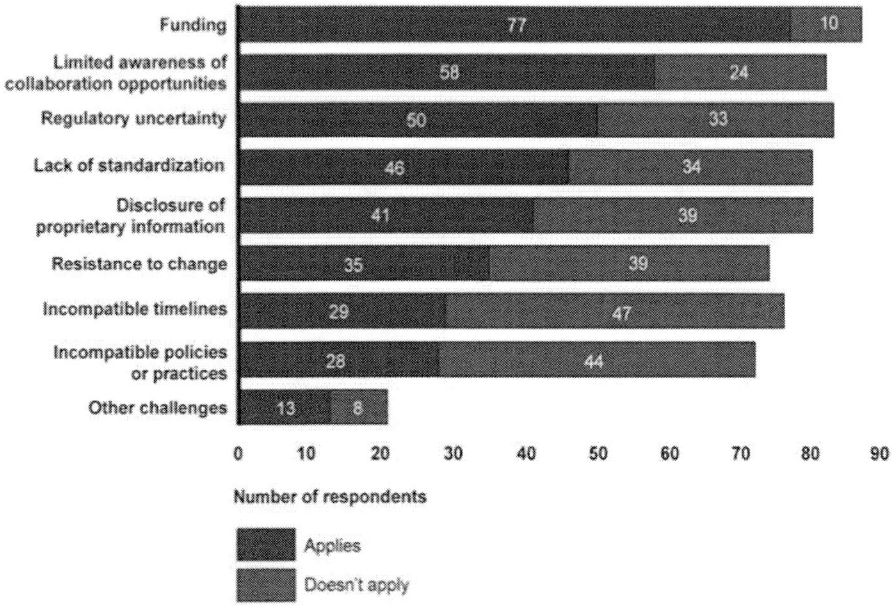

Source: GAO analysis of questionnaire of nonfederal stakeholders.

Note: Excludes respondents who selected "Don't know/Not sure" and those who did not check a response.

Figure 8. Challenges to Collaboration on Nanotechnology EHS Research, According to Questionnaire Respondents.

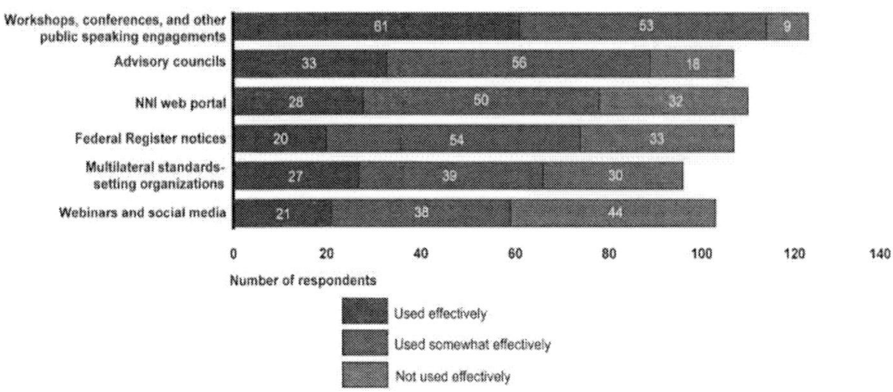

Source: GAO analysis of questionnaire of nonfederal stakeholders.

Note: Excludes respondents who selected "No basis to judge" and those who did not check a response.

Figure 9. Effectiveness of Mechanisms for Obtaining Input on the NNI Strategic Planning for EHS Research, According to Questionnaire Respondents.

When asked to rate the 2011 NNI EHS research strategy, most of those responding rated this strategy as somewhat or very effective at addressing nanotechnology EHS research needs. As seen in figure 10, while just over half of those responding to these questions indicated the 2008 NNI *Strategy for Nanotechnology-Related EHS Research* addressed these needs somewhat or very effectively, 79 out of 98 indicated the 2011 NNI EHS research strategy addressed these needs somewhat or very effectively.[53] Some of the respondents who commented in response to optional open-ended questions commended the 2011 NNI EHS research strategy for accurately capturing the input of the workshops' participants and summarizing the research needs, but others pointed to shortcomings, including a lack of prioritization and insufficient focus on implementation. For example, one respondent commented that the strategy would only be effective if it was implemented, and that it may not be sufficient to rely on the agencies' voluntary implementation efforts.

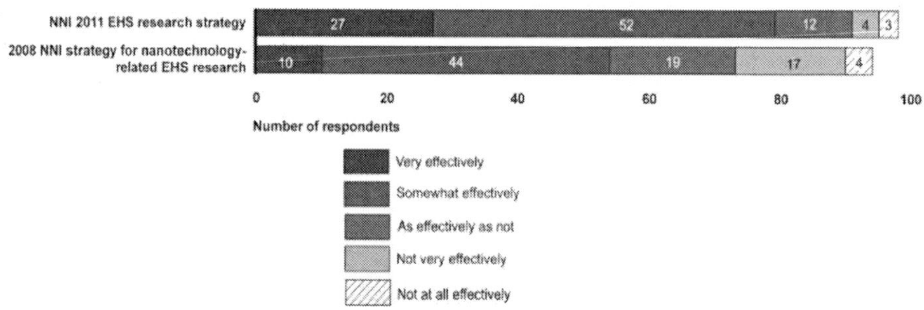

Source: GAO analysis of questionnaire of nonfederal stakeholders.
Note: Excludes respondents who selected "No basis to judge" and those who did not check a response.

Figure 10. Effectiveness of NNI EHS Research Strategies at Addressing Nanotechnology EHS Research Needs, According to Questionnaire Respondents.

Respondents Obtained Nanotechnology EHS Information from Various Sources

Additionally, respondents identified various sources, including the NNI member agencies, from which they obtained information on the potential EHS risks of nanotechnology. When asked to estimate the frequency with which they obtained nanotechnology EHS information from NNI member agencies in recent years, respondents reported obtaining information from the NNI member agencies in varying frequencies. More than half of the respondents to these questions reported occasionally or often obtaining information from EPA and NIOSH, as seen in figure 11.

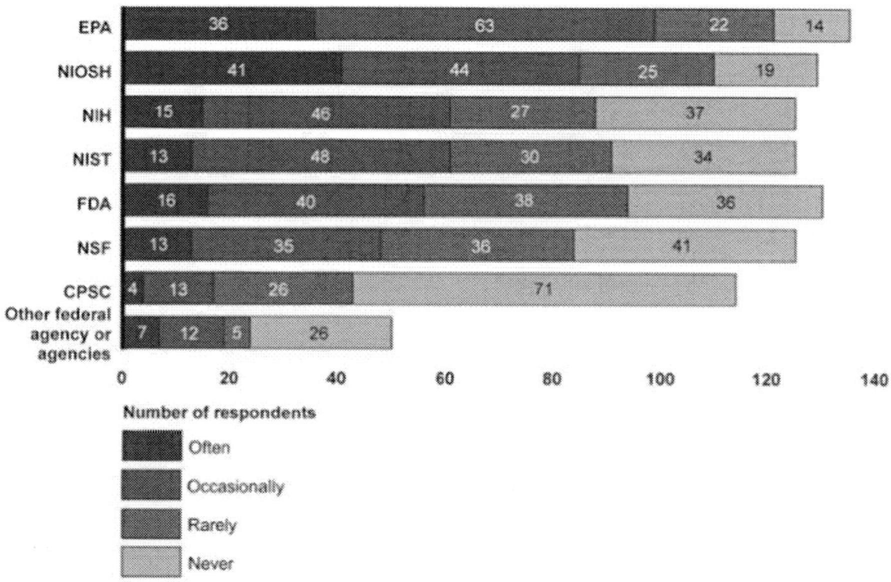

Source: GAO analysis of questionnaire of nonfederal stakeholders.

Note: Excludes respondents who selected "Don't know/Not sure" and those who did not check a response.

Figure 11. Frequency with Which Questionnaire Respondents Reported Obtaining Information on the Potential EHS Risks of Nanotechnology from NNI Member Agencies.

The 2011 NNI strategic plan calls for providing a "one-stop" Internet-based portal for nanotechnology information, including, among other things, scientific data such as characterization and toxicity measurements. About 80 percent of the 131 stakeholders responding to a question about how useful such a portal would be indicated that an Internet-based portal for nanotechnology EHS information would be generally or very useful to them. The NNI does not currently provide a portal that contains this information, but an NNCO official told us that the NNCO is implementing a web-based map that identifies the instrumentation and facilities for nanotechnology research at major universities and centers in the U.S. and federal labs.

When asked to rate the frequency with which they obtained information on the potential EHS risks of nanotechnology from nongovernmental sources, as seen in figure 12, more than two-thirds of respondents reported that they occasionally or often obtained information from each of the following four sources: (1) peer-reviewed scientific publications; (2) in-house research; (3) online databases; and (4) news outlets.

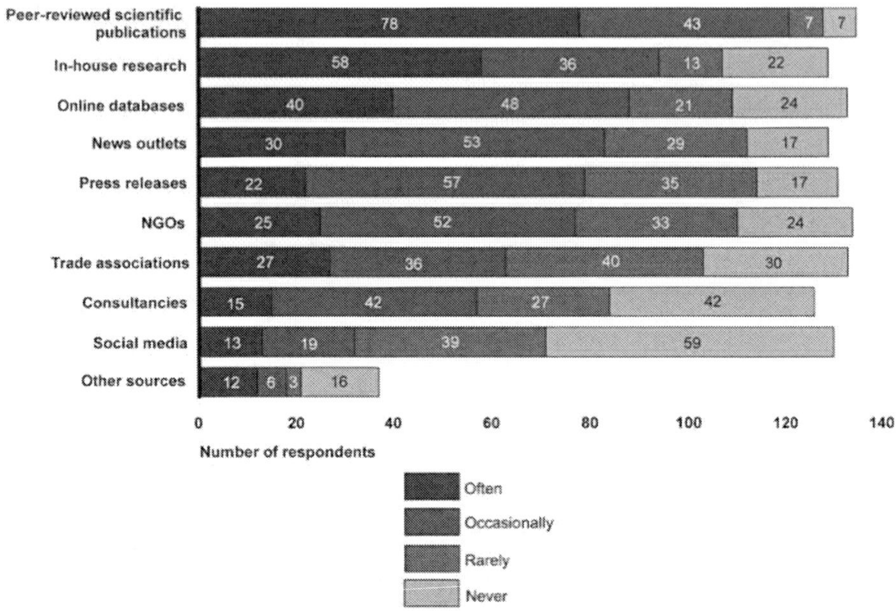

Source: GAO analysis of questionnaire of nonfederal stakeholders.

Note: Excludes respondents who selected "Don't know/Not sure" and those who did not check a response.

Figure 12. Frequency with Which Questionnaire Respondents Reported Obtaining Information on the Potential EHS Risks of Nanotechnology from Nongovernmental Sources.

NNI Strategy Documents Address or Partially Address Desirable Characteristics of National Strategies

NNI strategy documents address two and partially address four of the six desirable characteristics of effective national strategies that we identified in prior work and that offer policymakers and implementing agencies a management tool that can help ensure accountability and more effective results. The three NNI strategy documents—the 2011 NNI strategic plan, the 2011 NNI EHS research strategy, and the NNI Supplement to the President's 2012 Budget[54]—compose a national strategy for nanotechnology EHS research and create a framework for achieving NNI program goals and priorities.

We rated the NNI strategy documents to determine how well they jointly addressed the six characteristics that we have identified for effective national strategies, as described in table 3.

Table 3. Summary of Desirable Characteristics for a National Strategy

Desirable characteristic	Brief description
Purpose, scope, and methodology	Addresses why the strategy was produced, the scopeof its coverage, and the process by which it was developed.
Problem definition and risk assessment	Addresses the particular national problems and threats the strategy is directed toward.
Goals, subordinate objectives, activities, and performance measures	Addresses what the strategy is trying to achieve; steps to achieve those results; as well as the priorities, milestones, and performance measures to gauge results.
Resources, investments, and risk management	Addresses what the strategy will cost, the sources and types of resources and investments needed, and where resources and investments should be targeted by balancing risk reductions and costs.
Organizational roles, responsibilities, and coordination	Addresses who will be implementing the strategy, what their roles will be compared to others, and mechanisms for them to coordinate their efforts.
Integration and implementation	Addresses how a national strategy relates to other strategies' goals, objectives, and activities—and to subordinate levels of government and their plans to implement the strategy.

Source: GAO-04-408T.

Ideally, effective national strategies should fully address all of these characteristics. However, we recognize that by their nature, national strategies are intended to provide broad direction and guidance—rather than be prescriptive, detailed mandates—to the relevant implementing agencies. Thus it is unrealistic to expect all national strategies to provide details on each and every characteristic we identified. Moreover, the NNI member agencies have different statutory authority and functions that may significantly affect their research and research priorities. Nonetheless, we believe the more detail a strategy provides, the easier it is for the responsible parties to implement it and achieve its goals.

We reviewed the three NNI strategy documents for elements related to these characteristics, and based on the inclusion of these elements, rated how well the strategic documents address the six characteristics. According to our methodology, the strategy documents "address" a characteristic when they explicitly include all, or nearly all, elements of the characteristic and have sufficient specificity and detail. The documents "partially address" a characteristic when they include some or most of the elements of the characteristic with sufficient specificity and detail.[55] The documents "do not address" a characteristic when they do not include any elements of a characteristic or references are too vague or general to be useful. Additional details about our ratings and the elements that make up these characteristics are available in appendix I.

As described in table 4, we found, when reviewed as a whole, that the strategy documents address or partially address all of the desirable characteristics of a national strategy.

Table 4. Extent NNI Strategy Documents Address GAO's Desirable Characteristics with Respect to Nanotechnology EHS Research

Desirable characteristic	Address	Partially address	Do not address
1) Purpose, scope, and methodology	X		
2) Problem definition and risk assessment	X		
3) Goals, subordinate objectives, activities, and performance measures		X	
4) Resources, investments, and risk management		X	
5) Organizational roles, responsibilities, and coordination		X	
6) Integration and implementation		X	

Source: GAO analysis of the NNI strategy documents.

Purpose, Scope, and Methodology and Problem Definition and Risk Assessment (Desirable Characteristics 1 and 2)

Taken as a whole, the NNI strategy documents address the first two characteristics—purpose, scope, and methodology and problem definition and risk assessment—by including all, or nearly all, of their elements. These characteristics describe the scope of the strategy, describe how and why it was produced, define the problems the strategy intends to address, discuss causes and the operating environment, and provide a risk assessment or broad description of potential risks. Elements of these characteristics that are included in the documents are providing a clear statement of purpose, defining key terms, delineating what major functions or mission areas are covered, discussing problems with the current understanding of EHS implications of nanotechnology, and the quality of currently available data, among others.

Goals, Subordinate Objectives, Activities, and Performance Measures (Desirable Characteristic 3)

The NNI strategy documents partially address the third characteristic regarding goals, subordinate objectives, activities, and performance measures. This characteristic describes the overall desired results, hierarchy of strategic goals and subordinate objectives, priorities, milestones, outcome-related performance measures, and process for monitoring and reporting on progress, among others. The NNI strategy documents include a number of elements such as desired results, strategic goals, and subordinate objectives, but do not include, or do not fully develop, other elements such as priorities, milestones, or outcome-related performance measures. Performance information—such as outcome-related performance measures and milestones comprised of targets and time frames for meeting those measures—allows managers to identify performance problems and develop approaches that improve results. For the purposes of this report, we define performance information to mean data collected to measure progress toward achieving an established goal. A wide range of information can be relevant to program performance. Performance information can focus on various dimensions of performance such as outcomes, outputs, quality, timeliness, customer satisfaction, or efficiency. It can inform key management decisions such as setting program priorities, allocating resources, identifying program problems, and taking corrective action to solve those problems; or it can help determine progress in meeting the goals of programs or operations. The NNI strategy documents contain detailed research needs for nanotechnology EHS and report the annual funding of EHS research and the number of projects supported for selected years, but the documents do not prioritize among these needs or include outcome-related performance measures, targets, or time frames that allow for monitoring and reporting on progress toward meeting the research needs. Independent reviews of the prior NNI strategy documents also noted an absence of performance information. In 2010, the National Nanotechnology Advisory Panel recommended that the NNCO monitor metrics that assess, among other things, the NNI's progress on developing methodologies to assess plausible risks of nanotechnology. The advisory panel also recommended that the NEHI working group develop and implement a strategy that links EHS research activities with knowledge gaps and decision-making needs. The 2009 NRC review concluded that the 2008 NNI *Strategy for Nanotechnology-Related*

EHS Research could be an effective tool, but did not include a number of elements, including measures of research progress. We previously reported that, ideally, a national strategy would set clear desired results and priorities, outcome-related performance measures, and specific milestones while giving implementing parties flexibility to pursue and achieve those results within a reasonable time frame.[56] The NNI strategy documents give agencies wide latitude to develop research programs to meet the goals and research needs of the NNI, but we found that NNI member agencies vary in their identification and reporting of agency-specific performance information for nanotechnology EHS research that could align with the NNI research needs. The 2011 NNI EHS research strategy states that prioritization, timing, and staging of the research needs identified by the strategy are components of an implementation plan and should be developed within agency missions and appropriations. Similarly, the 2011 NNI strategic plan states that the document serves as a guide for individual agency implementation. Six of the seven NNI agencies we reviewed have documented performance measures, targets, or time frames for their EHS research. Three of the NNI member agencies—CPSC, FDA, and NIH—annually report EHSrelated performance information through their publicly available annual performance reports.[57] Three other NNI member agencies—EPA, NIOSH, and NIST—identify strategic goals and performance measures and targets in their program-level nanotechnology EHS strategies, but have not collected, or have not publically reported results of these performance measures.[58] NIOSH has previously published research summaries in its nanotechnology research progress reports. NIOSH officials told us that a forthcoming progress report will also document the results of performance measures identified in their nanotechnology EHS strategy. NSF reports planned EHS research activities as part of its annual budget request, but does not identify performance targets or measures for its nanotechnology EHS research beyond funding requested. According to an NNCO official, the NEHI working group is piloting an effort to gather information from the NNI member agencies to assess high-level progress toward meeting NNI research needs, but has not released the results of this work.

Resources, Investments, and Risk Management and Organizational Roles, Responsibilities, and Coordination (Desirable Characteristics 4 and 5)

The NNI strategy documents partially address the fourth and fifth characteristics describing resources, investments, and risk management and organizational roles, responsibilities, and coordination. We found that the documents include some, but not all, of the elements of these characteristics. With respect to the fourth characteristic, we found that the documents include elements relating to agency resources associated with the strategy such as descriptions of current activities of NNI member agencies and current investment levels. However, we found that the documents do not include, or do not fully develop, other elements such as the costs and types of resources—for example human capital or research and development costs—associated with implementing the strategy. An NNCO official told us that the NNI does not analyze the costs of resources to meet the EHS research needs because it does not control the funding for member agencies, and that each agency allocates funding and prioritizes its activities based on its mission area and available budget. According to this official, the 2011 NNI EHS research strategy identifies research goals, but it is up to the agencies to determine how their funding should be spent. Similarly, a NIST official told us that the NNI cannot dictate to agencies how or what research to fund. With respect to the fifth characteristic pertaining to organizational roles, responsibilities, and coordination, we found that the documents include elements such as lead, support, and partner roles for agencies and processes for coordination and collaboration, but do not include, or do not fully develop, other elements such as an accountability and oversight framework. The strategy documents include detailed descriptions of the NNI structure, describe coordination mechanisms such as the NEHI working group, and provide examples of collaborative activities related to EHS research. As we reported above, we found that NNI member agencies have engaged in numerous collaborative projects with federal and nonfederal partners and nonfederal partners generally rated these collaborations as generally or very useful. However, the strategy documents do not provide details on oversight of agency roles or how agencies will be held accountable to the goals and research needs of the NNI strategy documents.

Integration and Implementation (Desirable Characteristic 6)

The NNI strategy documents also partially address the sixth characteristic describing integration and implementation. This characteristic describes how the national strategy documents relate to subordinate levels of government and their plans to implement the strategy. As we have previously reported, to achieve a common outcome, collaborating agencies need to establish strategies that work in concert with those of their partners or are joint in nature.[59] We found that the NNI strategy documents include limited information regarding the elements of this characteristic. For example, the 2011 NNI EHS research strategy describes a framework for implementation, but does not provide detailed implementation guidance.[60] In addition, we found that the NNI strategy documents generally do not include information on the NNI member agencies' nanotechnology EHS strategies or plans or their integration with the national strategy. Five of the seven NNI member agencies we reviewed have developed such agency-level nanotechnology EHS research strategies. As noted above, EPA, NIOSH, and NIST have developed strategic plans specifically for EHS research. In addition, FDA has published a strategic plan for its regulatory science research efforts, which include nanotechnology, and developed a research plan specifically targeting nanotechnology.[61] Furthermore, NSF officials told us that the agency describes its strategic direction for nanotechnology, including current and proposed activities and funding by PCA, in its annual budget request.[62] CPSC includes nanotechnology as an external risk factor for its agencywide strategic plan, but the plan does not address EHS research directly. NIH officials told us they do not have agency or institute-specific strategic plans that specifically address nanotechnology EHS research, but nanotechnology EHS research plans developed by its institutes conform to the NNI strategy. Of the five agencies with nanotechnology EHS-related strategies, two—FDA and NIOSH—explicitly align agency research goals in their strategic plans for nanotechnology to NNI EHS research needs. The FDA and NIOSH strategic plans describe the agencies' EHS research goals, their framework for meeting those goals, and which NNI EHS research needs will be advanced. NIST's Nano-EHS program plan, EPA's Nanomaterial Research Strategy, and NSF's budget request to Congress describe agency research objectives. The documents do not explicitly identify NNI research needs; however, agency officials told us that they reflect the NNI strategic documents. The 2011 NNI strategic plan reports that NNI member agencies have initiated a mapping exercise to evaluate how this strategic plan relates to NNI member agencies' strategic plans and to the priorities of the administration. In addition, according to NNI member agency officials, individual agencies' implementation efforts of the 2011 NNI EHS research strategy are discussed at NEHI meetings. However, the results of this mapping exercise and proceedings of the NEHI are not publically available.

CONCLUSION

The increasing commercialization of nanotechnology-enabled products and information gaps related to nanotechnology EHS impacts underscore the importance of ensuring progress toward meeting EHS strategic goals and research needs. The NNI strategy documents help shape the policies, programs, priorities, resource allocations, and standards related to nanotechnology EHS research and activities. The NNI strategy documents address or partially address the desirable characteristics of effective national strategies, including overarching strategic goals and objectives such as the EHS research needs. However, the documents do not include some elements of the characteristics, which could make it more difficult for federal agencies and other stakeholders to implement the strategy and achieve the identified results. For example, progress toward achieving the strategy documents' goals and research needs cannot be tracked because the NNI does not report specific performance information, such as performance measures, targets, or time frames. In addition, the NNI strategy documents do not include estimates of the costs and types of resources associated with these goals and research needs.

Developing and coordinating performance information and cost estimates for the NNI's nanotechnology EHS research may be challenging because of the varied missions and priorities of the NNI member agencies. However, not having performance information that is aligned with the strategic goals and research needs of the NNI makes it difficult for agencies, policymakers, and stakeholders to determine the collective progress of the national nanotechnology program. Furthermore, developing performance information and including it in NNI strategy documents could help to strengthen two other desirable characteristics that the documents partially address. Specifically, this information could also support an accountability and oversight framework—that is currently not well-developed—by linking agency activities to specific measures of performance and creating consistent benchmarks to judge the overall progress of the NNI member agencies. In addition, developing this information could facilitate better integration between the NNI strategy documents and the NNI member agencies' nanotechnology EHS research strategies and goals. Providing cost estimates related to the NNI EHS research needs would allow the NNI member agencies, policymakers, and stakeholders to assess if investments are commensurate with costs of the identified needs. The NSTC, which is administered by the OSTP, does not direct funding to meet the NNI research needs; however, rough estimates of their costs could serve as guidance to the NNI member agencies as they individually determine their own budgets and priorities. In addition, the NSTC is well-positioned to develop these estimates due to its coordinating role across the member agencies of the NNI and corresponding access to expertise at these agencies.

Developing performance information and cost estimates could also support the analysis of the progress made toward achieving the goals and priorities established for the program, as required by the 21st Century Nanotechnology Research and Development Act. In addition, we are reiterating our 2008 recommendation that the Director of OSTP, in consultation with the Directors of the NNCO and OMB, provide better guidance to agencies on how to report their nanotechnology EHS research.[63] The continued absence of detailed guidance on how to report research under PCA 7 has contributed to the data reporting problems we identified. Therefore, agencies, policymakers, and stakeholders do not have access to accurate, consistent, and complete data on the federal government's investment in nanotechnology EHS research. Without reliable and up-to-date data, it will be difficult for agencies to accurately assess and report their progress toward their own performance measures, as well as the EHS research goals and needs identified in the 2011 NNI EHS research strategy.

Recommendations for Executive Action

To better offer policymakers and implementing agencies a management tool that can help ensure accountability and more effective results, we are making two recommendations to the Director of OSTP:

We recommend that the Director of OSTP coordinate development by the NNI member agencies of performance measures, targets, and time frames for nanotechnology EHS research that align with the research needs of the NNI, consistent with the agencies' respective statutory authorities, and include this information in publicly available reports. We also recommend that, to the extent possible, the Director of OSTP coordinate the development by the NNI member agencies of estimates of the costs and types of resources necessary to meet the EHS research needs.

Agency Comments and Our Evaluation

We provided a draft of this report to the Director, OSTP; Secretary, Commerce; Chairman, CPSC; Administrator, EPA; Secretary, Health and Human Services; and Director, NSF. OSTP and the agencies neither agreed nor disagreed with the recommendations. OSTP and Health and Human

Services provided technical and clarifying comments, which we incorporated as appropriate. The Department of Commerce, CPSC, EPA, and NSF indicated they had no comments on the report.

Sincerely yours,
Frank Rusco
Director,
Natural Resources and Environment

APPENDIX I: OBJECTIVES, SCOPE, AND METHODOLOGY

Our review examines (1) changes in federal funding for nanotechnology environmental, health, and safety (EHS) research from fiscal years 2006 to 2010; (2) the nanomaterials that National Nanotechnology Initiative (NNI) member agencies focused on in their EHS research in fiscal year 2010; (3) the extent to which NNI member agencies collaborate with stakeholders on nanotechnology EHS research and related strategies; and (4) the extent to which NNI strategy documents address desirable characteristics of national strategies. To conduct this work, we reviewed EHS research efforts funded by seven NNI member agencies, which collectively funded 93 percent of EHS research dollars in fiscal year 2010: the National Science Foundation (NSF), National Institutes of Health (NIH), Environmental Protection Agency (EPA), National Institute for Occupational Safety and Health (NIOSH), Food and Drug Administration (FDA), National Institute of Standards and Technology (NIST), and the Consumer Product Safety Commission (CPSC). The first six of these agencies represent the top six providers of EHS research funding from fiscal year 2006 to fiscal year 2010, and CPSC has an important role in ensuring the safe use of nanotechnology in consumer products.

To examine recent changes in federal funding for nanotechnology EHS research, we reviewed information published by the National Science and Technology Council (NSTC) in the NNI Supplements to the President's Budget. Specifically, we reviewed the actual agency investments reported in each program component area for fiscal years 2006 through 2010 for all NNI member agencies funding nanotechnology research.[1] For the dollar amounts that we adjusted for inflation, we used the Biomedical Research and Development Price Index to report funding in constant 2010 dollars. We consulted with the Office of Management and Budget (OMB) and officials at the seven selected agencies to determine the type of budget information reported as actual agency investments in these documents. For fiscal year 2010, the most recent year for which actual agency investment data were available, we also collected quantitative and qualitative project-level data (such as project funding amounts and project abstracts or progress reports) on all research projects that the seven selected agencies had categorized as Program Component Area (PCA) 7 (EHS research). The agencies do not report project-level data to OMB annually but report their total EHS research funding to OMB annually for inclusion in the NNI budget supplements published by NSTC. Therefore, we reviewed data on the individual projects included in the agencies' total EHS research funding in 2010. We provided the agencies with a spreadsheet template to use in listing their projects, which we based on the spreadsheet OMB had used in its data calls to collect project-level data from agencies on their 2006 and 2009 EHS research projects. Consistent with the approach used in OMB's data calls, we did not ask the agencies to use a standardized definition of what constitutes a "project;" instead, we deferred to each agency to identify their projects and consulted with agency officials as needed.

For some projects at NIH, NIOSH, and NIST, we grouped together some of the agencies' data entries and counted them together for the purposes of our analyses. Consequently, the numbers of projects we are reporting may not match the number reported by these agencies elsewhere. Specifically, for NIH, we grouped the 14 supplement grants together with their main grants and counted each of those groups as a single project. We also grouped together the individual subprojects of each of NIH's multiproject "parent" grants reported as EHS and counted each of those groups as a single project, for a total of seven groups of multiproject grants. In addition, we grouped together two grants made by different NIH institutes for the same project. For NIOSH, we identified two project entries with identical project descriptions. After consulting with NIOSH officials, we grouped those two entries together and counted them as a single project. For NIST, the agency reported a separate entry for each of its Project Tracking Numbers and explained that a single technical project is often associated with more than one Project Tracking Number. NIST officials instructed us to group together the entries with identical titles and count each of those groups as a single project.

To assess the reliability of the agencies' fiscal year 2010 data, we sent each of the seven selected agencies a set of questions regarding the data and the information systems used to produce and store them. We also reviewed related supporting documentation, such as user manuals and data dictionaries for information systems, and, for some projects, copies of the associated interagency agreements. We examined the data for obvious errors, compared each agency's total fiscal year 2010 PCA 7 funding reported to us to the totals reported in the NNI Supplement to the President's 2012 Budget, and consulted with agency officials to understand the reasons for any differences. We determined that the data were sufficiently reliable for the purposes of our analyses. However, we did not attempt to verify the funding amounts reported for each project— for example, by reviewing documentation of agencies' actual expenditures for each project. We reviewed the qualitative project data, including abstracts and project reports, to assess whether the agencies had appropriately categorized the research projects they reported as EHS research—that is, whether the projects met the definition of program component area (PCA) 7. As described in this report, the definition of PCA 7—research primarily directed at understanding the EHS impacts of nanotechnology development and corresponding risk assessment, risk management, and methods for risk mitigation—is the only written guidance available to agencies on how to report their nanotechnology research in this program component area. Determining whether research meets this definition is an inherently subjective process. Therefore, rather than definitively conclude that any projects did not meet the definition, we assigned projects a designation of "not clearly PCA 7" if we were ultimately unable to identify any EHS research, or it was not clear that EHS was the primary focus of the research. For example, some projects studied various nanotechnology-enabled drugs, but it was not clear to what extent the research was directed at the safety of the drugs versus their efficacy. Some other projects appeared to be closely related to program component areas other than PCA 7, such as PCA 4 (instrumentation research, metrology, and standards for nanotechnology) or PCA 8 (education and societal dimensions), and it was not clear that PCA 7 was the most appropriate category. To minimize bias and to ensure the consistency of our evaluation, two analysts independently analyzed the project data and used their professional judgment to assess whether each project met the PCA 7 definition. The two analysts discussed those projects where they did not agree on whether the research was primarily directed at EHS and reached agreement. For categorization of projects that appeared questionable to us, we generally asked agencies to explain why they had reported those projects as EHS research. We reviewed agencies' responses and modified our determinations for projects as appropriate given the additional support provided by the agencies. For two projects funded by NIST and 18 by NSF, the agencies had reported only a portion of the total project funding under PCA 7. In those instances, we assessed whether at least some of the project was primarily directed at EHS. However, we did not verify that the funding amounts reported as EHS research were correct because that information is, according to these agencies' officials, based on the discretion of the knowledgeable agency program staff. We also reviewed our 2008 report in this area, which found that 22 of 119 fiscal year 2006 projects reported as EHS research were miscategorized, and recommended that the Director of the Office of Science and Technology Policy (OSTP), in consultation with the Directors of the National Nanotechnology Coordinating Office and OMB, provide better guidance to agencies on how to report nanotechnology EHS research.[2] We asked OMB and OSTP to identify any guidance provided to agencies on how to report their EHS research and to describe any steps taken since our prior report to improve such guidance. To identify the nanomaterials that the seven selected NNI member agencies focused on in their EHS research in fiscal year 2010, we reviewed qualitative project data—such as abstracts and progress reports—for those projects we determined were primarily directed at EHS. We also consulted with agency officials. One analyst identified, and a second analyst verified, the nanomaterials in the project documentation. We attempted to identify the nanomaterials that the projects actually studied in fiscal year 2010. However, some projects are multiyear efforts. The level of detail in the documentation we reviewed for each project varied, and the documentation did not always indicate which materials were studied in which years. For the purposes of our analysis, we assumed that all nanomaterials identified for a project were studied in 2010, unless the documentation specified otherwise. It is possible that some additional nanomaterials were studied by these projects but not referenced in the documents we reviewed. It is also possible that some of the nanomaterials referenced in project documents were not actually studied until later years of a project. However, we do not have any reasons to suspect that such variances would be significant or would substantially change our findings.

We assigned each particular nanomaterial identified to one of five categories: carbon-based nanomaterials; metal-based nanomaterials; semiconductor-based nanomaterials; organic nanomaterials; and other nanomaterials. In addition, some projects were not targeted to particular nanomaterials. An analyst with previous academic and professional experience in nanotechnology helped identify the categories we used and verified that the materials were placed in the appropriate categories for each project. After reviewing the fiscal year 2010 EHS research by the selected agencies, we selected the five categories of nanomaterials we used, based on the primary composition of the materials, to encompass the range of materials identified in the projects. To inform our selection of the categories we used for grouping nanomaterials, we reviewed literature related to nanotechnology EHS issues from federal agencies (NIH and EPA), nonprofit organizations (the Woodrow Wilson International Center for Scholars' Project on Emerging Nanotechnologies and the World Technology Evaluation Center, Inc.), and a market research firm that tracks nanotechnology (Lux Research). We found that various classes or categories were used to present information on nanomaterials. For materials identified as composites, we assigned them to categories based on the primary nanomaterials from which they were formed—for example, we assigned carbon nanotube nanocomposites to the carbon-based nanomaterials category. To determine the extent to which the NNI member agencies collaborate with stakeholders on nanotechnology EHS research and related strategies, we (1) discussed with agency officials how their agencies collaborate on nanotechnology EHS research and NNI's role in facilitating that collaboration, and (2) we obtained documentation on these collaborative efforts. We conducted a review of formal collaborative efforts (i.e., those that are documented in written agreements) that focused on nanotechnology EHS research initiated from February 2008 to October 2011 while the 2008 NNI *Strategy for Nanotechnology-Related EHS Research* was in effect.[3] We chose this time frame because this strategy—the first NNI strategy to focus on EHS research—established collaboration as necessary for its implementation efforts. The strategy was in effect for a number of years, which allowed us to get a more complete picture of the number of collaborative projects completed during its implementation. We only included collaborative efforts that were formalized by written agreements because those provided a clearer description of the agencies' contributions of resources. We obtained an initial list of such efforts from published NNI sources as well as an OMB data call used to compile EHS-related nanotechnology projects for the 2011 NNI EHS research strategy. We sent our initial list to the selected NNI agencies for their review and revision, and analyzed supporting documents provided by the agencies. We provide relevant information from the supporting documents for each of the projects within our scope in appendix II.

We administered a web-based questionnaire to a nonprobability sample of nonfederal stakeholders,[4] including those affiliated with academia, companies, nongovernmental organizations (NGO), and state and local governments, to obtain their views on collaboration with the NNI member agencies on EHS research and the EHS research strategies. Because the population of nonfederal stakeholders is diverse, wide-ranging, and difficult to reliably define, we chose to use a nonprobability sample of individuals who (1) had expertise in the field of nanotechnology, and (2) had interacted with the NNI in the past few years, or who were representatives of organizations and companies suggested to us through our scoping interviews. We identified relevant potential respondents from several sources, including the list of participants in NNI workshops that were used to solicit input for the 2011 NNI EHS research strategy[5] and participants in a review of the NNI by the President's Council of Advisors on Science and Technology. We also included members of nanotechnology-related trade associations and other organizations through interviews with nanotechnology experts using an iterative process, often referred to as "snowball sampling." We excluded federal employees and contractors, individuals or organizations that are not located in the United States, and those for whom we could not obtain contact information. In addition, individuals who were not considered to have expertise in the nanotechnology field based on their job titles were excluded.[6] These sources provided us with a list of 228 stakeholders chosen as potential respondents. Of these 228 individuals, five subsequently were excluded from our sample because they had invalid email addresses, were duplicates, or were not available for an extended period, giving us a final sample of 223 potential respondents. We categorized the population surveyed by affiliation type, usually based on information provided on each entity's website, as follows: (1) "academic," which includes universities or university-affiliated programs or centers; (2) "company," which includes companies, corporations, or other for-profit entities whose primary purpose is to sell products, services, or information; (3) "organization," which includes NGOs or other entities whose primary purpose is public interest or mutual benefit; and (4) "state and local government." The questions included in the questionnaire were developed using issues raised in our scoping interviews and background literature search. Most of our questions provided respondents with answer options identified through our interviews and background literature searches. In certain cases, we also provided respondents with an "other" category which allowed them to write in additional responses we did not identify. However, because few respondents chose this option, we presented only some of these responses as illustrative examples. The questionnaire went through internal reviews, including independent GAO survey experts, as well as four pretests with two stakeholders who work for companies, one academic, and one stakeholder from an NGO. We chose the pretesters to reflect the diversity of affiliations of the nanotechnology experts included among our respondents. We revised and clarified the questions and introductory material to the questionnaire based on comments obtained in the internal reviews and from the pretesters. We administered the questionnaire through a secure server. When we completed the final survey questions and format, we sent an e-mail announcement of the questionnaire to the nonfederal stakeholders in our sample on September 23, 2011. Stakeholders were notified that the questionnaire was available online and were given unique passwords and user names on October 11, 2011. We sent follow-up e-mail messages on October 18, 2011, October 25, 2011, and November 2, 2011 to those who had not yet responded. Then we contacted all nonrespondents by telephone, starting on October 21, 2011. The questionnaire was available online until December 6, 2011. Because this was not a sample survey, it has no sampling errors. However, the practical difficulties of conducting any survey may introduce errors, commonly referred to as nonsampling errors. For example, difficulties in interpreting a particular question, sources of information available to respondents, or entering data into a database or analyzing them can introduce unwanted variability into the survey results. We took steps in developing the questionnaire, collecting the data, and analyzing them to minimize such nonsampling errors. For example, we worked with social science survey specialists to design the questionnaire. We pretested the draft questionnaire with four nonfederal stakeholders to ensure that the questions were relevant, clearly stated, and easy to understand. When we analyzed the data, an independent analyst checked all computer programs. Since this was a web-based questionnaire, respondents entered their answers directly into the electronic questionnaire, eliminating the need to key data into a database, minimizing error. Of the 223 potential respondents, 138 completed the questionnaire during our time frame, for an overall response rate of about 62 percent.[7] The numbers responding by affiliation type as categorized by us were (1) "academic"—42 out of 62; (2) "company"—65 out of 114; (3) "organization"—24 out of 39; and (4) "state and local government"— 7 out of 8. To determine the extent the NNI strategy documents address desirable characteristics of national strategies, we compared three key NNI strategic documents 8 against criteria for desirable characteristics of national strategies. Prior GAO reports have identified six desirable characteristics for national strategy documents that would help shape the policies, programs, priorities, resource allocations, and standards that would enable federal agencies and other stakeholders to implement the strategies and achieve the identified results.[9] National strategies that address these characteristics offer policymakers and implementing agencies a management tool that can help ensure accountability and more effective results.[10] Table 5 provides the desirable characteristics and the elements we looked for.

Table 5. Elements of Desirable Characteristics of National Strategies

Desirable characteristic	Brief description	Elements of characteristic
Purpose, scope, and methodology	Addresses why the strategy was produced, the scope of its coverage, and the process by which it was developed.	•Statement of broad or narrow purpose, as appropriate. •How it compares and contrasts with other national strategies. •What major functions, mission areas, or activities it covers. •Principles or theories that guided its development. •Impetus for strategy (e.g., statutory requirement or event). •Process to produce strategy (e.g., interagency task force; state, local, or private input). •Definition of key terms.
Problem definition and risk assessment	Addresses the particular national problems and threats the strategy is directed toward.	•Discussion or definition of problems, their causes, and operating environment. •Risk assessment, including an analysis of threats andvulnerabilities. •Quality of data available (e.g., constraints, deficiencies, and "unknowns").
Goals, subordinate objectives, activities, and performance measures	Addresses what the strategy is trying to achieve, steps to achieve those results, as well as the priorities, milestones, and performance measures to gauge results.	•Overall results desired (i.e., "end-state"). •Hierarchy of strategic goals and subordinate objectives. •Specific activities to achieve results. •Priorities, milestones, and outcome-related performance measures. •Specific performance measures. •Process for monitoring and reporting on progress. •Limitations on progress indicators.
Resources, investments, and risk management	Addresses what the strategy will cost, the sources and types of resources and investments needed, and where resources and investments should be targeted by balancing risk reductions and costs.	•Resources and investments associated with the strategy. •Types of resources required, such as budgetary, human capital, information technology, research and development, contracts. •Sources of resources (e.g., federal, state, local, and private). •Economic principles, such as balancing benefits and costs. •Resource allocation mechanisms, such as grants, in-kind services, loans, or user fees. •"Tools of government" (e.g., mandates or incentives to spur action).

Desirable characteristic	Brief description	Elements of characteristic
		•Importance of fiscal discipline. •Linkage to other resource documents (e.g., federal budget). •Risk management principles.
Organizational roles, responsibilities, and coordination	Addresses who will be implementing the strategy, what their roles will be compared to others, and mechanisms for them to coordinate their efforts.	•Roles and responsibilities of specific federal agencies, departments, or offices. •Roles and responsibilities of state, local, private, and international sectors. •Lead, support, and partner roles and responsibilities. •Accountability and oversight framework. •Potential changes to current organizational structure. •Specific processes for coordination and collaboration. •How conflicts will be resolved.
Integration and implementation	Addresses how a national strategy relates to other strategies' goals, objectives, and activities—and to subordinate levels of government and their plans to implement the strategy.	•Integration with other national strategies (horizontal). •Integration with relevant documents from implementing organizations (vertical). •Details on specific federal, state, local, or private strategies and plans. •Implementation guidance. •Details on subordinate strategies and plans for implementation (e.g., human capital, and enterprise architecture).

Source: GAO-04-408T.

To assess whether the documents addressed these desirable characteristics, two analysts independently reviewed the three NNI strategy documents for the elements of each characteristic. We examined the documents for inclusion of each element and for sufficient specificity and detail. The two analysts compared their assessments, discussed any discrepancies, and agreed upon a determination for each element. Each characteristic was then given a rating of "address," "partially address," or "does not address." According to our methodology, the strategy documents "address" a characteristic when it explicitly includes all, or nearly all, elements of the characteristic and has sufficient specificity and detail. The documents "partially address" a characteristic when it includes some or most of the elements with sufficient specificity and detail. The "partially address" category includes a range that varies from explicitly including most of the elements to including as few as one of the elements of a characteristic. A strategy "does not address" a characteristic when it does not include any elements of a characteristic or references are too vague or general to be useful. Because the three NNI documents have different purposes and audiences, an individual document need not address each element of each desirable national strategy characteristic. Therefore, we reviewed the three NNI strategy documents as a whole. In the case of the 2011 NNI strategic plan and Supplement to the President's 2012 Budget, we focused on sections related to EHS research, where applicable. We identified agency documents related to nanotechnology EHS research strategies and performance. Documents were identified based on interviews with agency officials, review of agency websites, and a data collection instrument sent to agencies requesting any documentation related to the planning or performance of their nanotechnology EHS research programs. In order to assess the extent of linkage between agencies' nanotechnology EHS research and national priorities of the NNI, we evaluated agency management documents for explicit referrals to NNI activities or guidance, including strategic documents such as the NNI strategic plan or EHS research strategy, or activities of the NNI's planning and coordinating bodies. We also reviewed agencies' Performance Reports in order to assess agencies' use of performance information related to nanotechnology EHS research efforts. We conducted this performance audit from February 2011 through May 2012 in accordance with generally accepted government auditing standards. Those standards require that we plan and perform the audit to obtain sufficient, appropriate evidence to provide a reasonable basis for our findings and conclusions based on our audit objectives. We believe that the evidence obtained provides a reasonable basis for our findings and conclusions based on our audit objectives.

APPENDIX II: COLLABORATIVE NANOTECHNOLOGY ENVIRONMENTAL, HEALTH, AND SAFETY RESEARCH AGREEMENTS

Table 6 identifies collaborative nanotechnology EHS research project agreements initiated by the NNI member agencies included in our review from February 2008 to October 2011. The table identifies only those projects specifically focused on nanotechnology. All the information in the table is based on the written agreements, which were provided by NNI member agencies' officials.

Table 6. Collaborative Nanotechnology Environmental, Health, and Safety Research Agreements

Participating federal agencies	Nonfederal partners	Date initiated	Type of collaboration	Title	Purpose
Consumer Product Safety Commission (CPSC), National Institute of Standards and Technology (NIST)	NA	April 2008	Memorandum of understanding	Memorandum of understanding between CPSC and NIST's Building and Fire Research Laboratory	Develop best practices to identify the presence of and characterize and quantify potential release of nanomaterials from flame retardant consumer products, to help assess the safety of these products. The cooperating agencies plan to work closely to expedite the development and availability of methods for the characterization and quantification of the release of nanomaterials from a variety of products (e.g., textiles, plastics) which are assessed using any of a number of standard fire tests.
CPSC, National Institute for Occupational Safety and Health (NIOSH)	NA	July 2008	Interagency agreement	Nano product exposure assessment	Conduct laboratory investigations on the emissions of nanoscale titanium dioxide from selected consumer products.

Table 6. (Continued)

Participating federal agencies	Nonfederal partners	Date initiated	Type of collaboration	Title	Purpose
					The project also is to involve developing the appropriate sampling and analysis methods for nanomaterials incorporated into selected consumer products. The first products to be investigated are bathroom cleaners that utilize titanium dioxide as a catalyst to destroy bacteria or to break down organic material. These products are believed to be used by consumers and in occupational settings (e.g., hotel cleaning). CPSC staff interests involve emissions during consumer use while NIOSH is concerned with emissions that occur in occupational use scenarios.
CPSC, NIST	NA	Initiated July 2009, extended September 2009	Interagency agreement	Exposure and fire hazard assessment of nanoparticles in fire safe consumer products	Determine if nanoparticles that are traditionally used to improve the fire performance of foam and barrier fabricsare released during simulated normal conditions (i.e., mechanical compression and saliva).

Participating federal agencies	Nonfederal partners	Date initiated	Type of collaboration	Title	Purpose
					In addition, determine if the release of nanoparticles also creates a decrease in the fire performance of these components.
CPSC, NIOSH	NA	March 2010	Interagency agreement	Pulmonary effects of titanium dioxide nanoparticles released from aerosol spray products	Evaluate the acute bioactivity of a particulate aerosol from a bathroom cleaner or sanitizer containing titanium dioxide nanoparticles by exposing rats by inhalation to the aerosolized spray particles (low and high dose) for 2hours and monitor pulmonary responses at 1, 7, and 28 days post-exposure.
CPSC, NIOSH	NA	April 2010	Interagency agreement	Exposure assessment of silver nanoparticles in treated products	Use scientifically credible protocols to evaluate exposure potential to nanosilver from related products. NIOSH plans to test the broad hypothesis that nanosilver is released from products under conditions ofintended use.
CPSC, EPA	NA	May 2010	Interagency agreement	Exposure assessment of silver nanoparticles in select children's consumer products	Evaluate children's potential exposures to nanosilver from consumer products marketed exclusively for children (e.g., pacifiers, plush toys)

Table 6. (Continued)

Participating federal agencies	Nonfederal partners	Date initiated	Type of collaboration	Title	Purpose
					using scientifically credible protocols. EPA plans to test the broad hypothesis that nanosilver is released from children's consumer products under conditions of intended use.
CPSC, NIOSH	NA	March 2011	Interagency agreement	Simulated worker exposure to silver from use of nano-enabled consumer products	Use scientifically defensible methods to evaluate the release of silver from consumer products and model exposure potential. NIOSH plans to test the broad hypothesis that silver is released from nanosilver enabled products under conditions of intended use in the work environment.
CPSC, NIOSH	NA	May 2011	Interagency agreement	Exposure assessment and pulmonary effects of silver nanoparticles released from aerosol spray products	Characterize the generated particulate from a product containing silver nanoparticles aerosol including the mass, particle size distribution, and chemical composition of the particles.

Participating federal agencies	Nonfederal partners	Date initiated	Type of collaboration	Title	Purpose
					NIOSH staff plan to exposerats by inhalation to the aerosolized spray particles (low and high dose) for 2 hours and monitor pulmonary responses at 1, 7, and 28 days post-exposure.
CPSC, NIST	NA	May 2011	Interagency agreement	Exposure assessment of carbon nanotubes in sports equipment	Determine if nanomaterials that are reported to be used in selected sports equipment are present in these products and are released during simulated use conditions. In addition, this project seeks to determine if the release of nanoparticles improve the performance ofthese products.
CPSC, NIST	NA	July 2011	Interagency agreement	Quantify and characterize released nanoparticles from thermoset and thermoplastic samples	Measure the impact of nanotechnologies on the flammability of the thermoplastic- and thermoset-based consumer products and develop risk models for human exposure to materials released from these consumer products.

Table 6. (Continued)

Participating federal agencies	Nonfederal partners	Date initiated	Type of collaboration	Title	Purpose
CPSC, NIST	NA	August 2011	Interagency agreement	Characterization of airborne nanoparticles released from consumer products	Develop testing and measurement protocols for determining the quantities and properties of nanoparticles released from flooring finishes and interior paints, and their subsequent airborne concentrations. Develop essential methods and data to assess the release as well as accumulation of nanoparticles at the surfaces of these products that seek to assist in estimating occupant exposure and developing strategiesto manage and mitigate these exposures.
CPSC, Environmental Protection Agency (EPA)a	NA	September 2011	Interagency agreement	Risk assessment for manufactured nanoparticles used in consumer products	Determine whether the nanoparticles exposures to humans and organisms resulting from consumer product and environmental exposures will substantially differ in both physicochemical and toxicological properties from nanoparticles that are

Participating federal agencies	Nonfederal partners	Date initiated	Type of collaboration	Title	Purpose
					newly synthesized in the laboratory or incorporated directly into a product. These differences may have significant consequences with respect to nanoparticles bioavailability and other toxicity measures. This study plans to test these measures in *in vitro*and *in vivo*experiments, and produce mechanism-based results relating toxic effects to nanoparticles physicochemical properties.
CPSC, National Science Foundation (NSF)	NA	September 2011	Interagency agreement	Interagency agreement between the CPSC and NSF	Develop innovative tools for measuring the potential health impact of nanotechnologies used in the flammability treatment of thermoplastic- and thermoset-based consumer products and develop risk models for human exposure to materials released from these consumer products.

Table 6. (Continued)

Participating federal agencies	Nonfederal partners	Date initiated	Type of collaboration	Title	Purpose
EPA	United Kingdom (UK) Natural Environment Research Council, UK Engineering and Physical Sciences Research Council, UK Department of Environment, Food, and Rural Affairs, UK Environment Agency	March 2009	Memorandum of understanding	Memorandum of understanding for a joint request for applications for nanotechnology research proposals	Establish a cooperative framework for the EPA, the UK Natural Environment Research Council, the UK Engineering and Physical Sciences Research Council, the UK Department of Environment, Food, and Rural Affairs, and the UK Environment Agency to organize, plan, and execute a joint Request for Applications for research related to the environmental and health implications of released nanomaterials.
EPA, NSF, U.S. Department of Agriculture's National Institute of Food and Agriculture	NA	Solicitation issued November 2009	Grant or cooperative agreement	Increasing scientific data on the fate, transport, and behavior of engineered nanomaterials in selected environmental and biological matrices	Increase scientific knowledge on the partitioning of nanomaterials in various media and increase data on movement and transformation capacities. Address the urgent needs to scientifically understand the fate and properties of nanoscale materials and additives that may be used or introduced into foods.

Participating federal agencies	Nonfederal partners	Date initiated	Type of collaboration	Title	Purpose
					Assess the adequacy of the existing characterization methods to study the critical questions, and establish the baseline for the needs of new characterization methodology. Provide guidance on the extent of future investigation needs on the nanoscale food materials and additives.
EPA, NIST	NA	July 2011	Interagency agreement	Measurements to support research in nanoparticles	Investigate aggregation of engineered nanoparticles in an artifact of air sampling on filter media and investigate how engineered nanoparticles change microscopically with age in the atmosphere.
EPA	International Life Sciences Institute Research Foundationb	September 2011	Contract	Support for better understanding the release of nanomaterials to the environment workshop (Phase 2)	Evaluate the “state of the science” for release measurement for multi-walled carbon nanotubes in polymer. The project seeks to use information (methods,studies, guidance, etc.) collected in response to a data call and literature search and provide that information to groups of subject matter experts selected by the steering committee.

Table 6. (Continued)

Participating federal agencies	Nonfederal partners	Date initiated	Type of collaboration	Title	Purpose
					This information is to also be made available in the project's information catalogue to the extent it does not reveal confidential business information.
Food and Drug Administration (FDA), National Institutes of Health's (NIH) National Cancer Institute (NCI), and NIST	NA	July 2009	Memorandum of understanding	Memorandum of understanding between the FDA, NCI, and NIST for the Nanotechnology Characterization Laboratory (NCL) and related nanotechnology activities	Facilitate the development of nanotechnologies that constitute novel research tools and safer, more effective cancer therapies by establishing a framework for effective risk identification, assessment, and evaluation of emerging products based on nanotechnology. This collaboration is to be focused primarily on the NCL and directly related activities.
FDA's Center for Biologics Evaluation and Research, NCL, NIST	Science Applications International Corporation (SAIC)-Fredrickc	February 2011	Grant mechanism or other opportunities	*In vitro*evaluation of effects of engineered nanomaterials on blood platelets	Develop a panel of *in vitro*assays for evaluation of effects of nanomaterials on blood platelets.

Participating federal agencies	Nonfederal partners	Date initiated	Type of collaboration	Title	Purpose
					Research validation of the assays for different representative typesof nanomaterials, soluble nanoparticles, and nanoobjects in suspension—representative metal, carbon, polymer, dendrimer, liposome nanoparticles, nanoparticles containing composites, or nanostructured macrosurfaces. Investigate a structure-function relationship of a nanoparticle-platelet interaction in a model of different size-charge modified poly(amidoamine) dendrimers.
FDA's National Center for Toxicological Research, Center for Food Safety and Applied Nutrition, Center for Veterinary Medicine; National Toxicology Program; Pacific Northwest National Laboratories	NA	February 2011	Grant mechanism or other opportunities	Development and evaluation of exposure dosimetry methods to optimize the standard *in vitro*mammalian genotoxicity assays for assessing engineered nanomaterials	Evaluate whether the existing *in vitro*mammalian genotoxicity assay is suitable for assessing the genotoxicity of nanomaterials. Explore the possible mechanisms underlying genotoxicity of engineered nanomaterials by conducting genomic analysis.

Table 6. (Continued)

Participating federal agencies	Nonfederal partners	Date initiated	Type of collaboration	Title	Purpose
					Identify potential improvements to the assay and general strategies for evaluating nanomaterials. Examine whether the suitable methods and other experiences learned from the micronucleus assay are applicable to other genotoxicity tests, such as mouse lymphoma assay and *in vivo*micronucleus assay.
FDA, NCI, NCL	SAIC-Fredrickc	July, 2011	Interagency agreement	*In vitro* dermal penetration and *in vivo*distribution of nanoparticles used in food and cosmetics	Utilize free and surface model nanoparticles (such as dendrimers, nano-silver or nano-gold) with the size range typically found in food and cosmetics to assess their characteristics and correlate the effect of surface modification on the ability of nanomaterials to penetrate various tissue barriers in different *in vitro*and *in vivo*systems.

Participating federal agencies	Nonfederal partners	Date initiated	Type of collaboration	Title	Purpose
					Supply information related to how nanomaterial physiochemical properties affect the ability of these nanoscale materials to interact with various biological systems and, as a result, to evaluate the benefits versus the risks when making regulatory approval decisions in order to facilitate the development and commercialization of safe nano-based cosmetics, foods, and other FDA-regulated products.
FDA's Center for Food Safety and Applied Nutrition and National Center for Toxicological Research	Key Laboratory for Biomedical Effects of Nanomaterials and Nanosafety, Chinese Academy of Sciences	Fiscal year 2011	Grant mechanism or other opportunities	Use of electron spin resonance spectroscopy and biomarkers of oxidative damage to assess the safety of nanoscale materials used in cosmetics	Develop rapid and predictive tests to assess toxicity elicited by nanosized materials. The tests may be used by the agency and the regulated industry to identify potentially hazardous nanosized materials in cosmetic products.
NIH's National Institute of Environmental Health Sciences (NIEHS), NIOSH	NA	Initiated in June 2008 and extended in June 2009 and March 2010	Interagency agreement	Comprehensive toxicological assessment of occupationally-relevant exposures	Characterize workplace exposure to selected engineered nanoparticles. This proposal is to focus on addressing the lack of exposure data.

Table 6. (Continued)

Participating federal agencies	Nonfederal partners	Date initiated	Type of collaboration	Title	Purpose
					Twelve field investigations at sites where engineered nanoparticles are handled and manufactured are to be conducted. The focus of these investigations is to be nanoparticles of interest to NIEHS and NIOSH.
NIEHS, NIOSH	NA	Initiated in February 2009 and extended in December 2009	Interagency agreement	Cardiovascular toxicity assessment of subchronic inhalation exposure to fullerene C60	Evaluate potential cardiovascular toxicity of subchronic inhalation exposure to fullerene C60 in animal models using molecular and biochemical analysis of cardiovascular tissue and blood samples. The findings are to be correlated with the histopathological, particle distribution, and blood chemistry results being obtained in other National Toxicology Program studies of this proposal.

Participating federal agencies	Nonfederal partners	Date initiated	Type of collaboration	Title	Purpose
NCI, NIH's National Institute of Biomedical Imaging and Bioengineering, NIEHS	NA	Initiated in April 2010 with NIEHS; NCI joined in September 2010	Direct citation agreement	Nano registry	Establish an authoritative nanomaterial registry, whose primary function is to provide consistent and authoritative information on the biological and environmental interactions of well-characterized nanomaterials, as well as links to associated publications, modeling tools, computational results and manufacturing guidance. The registry seeks to provide acurated data source and assessment information on the health and environmental effects of well-characterized nanomaterials. A set of minimal information about nanomaterials, ontology, and standards are to be developed through a community effort with broad representation to establish the authoritativeness of the registry.

Table 6. (Continued)

Participating federal agencies	Nonfederal partners	Date initiated	Type of collaboration	Title	Purpose
NCI, NCL, NIEHS	SAIC-Fredrickc	April 2010	Direct citation agreement	NIEHS nanomaterial characterization and informatics	NIEHS requests NCI and SAIC-Fredrick support for physiochemical characterization of engineered nanomaterials. NCL seeks to provide data to support and compare in-house characterization from program grantees. Facilitate the dissemination of findings and provide maximum impact to the field by providing direct access to a database for inclusion of nanomaterials physiochemical characterization and biological responsedata.
NIEHS, NIOSH	NA	November 2010	Interagency agreement	Comprehensive toxicological assessment of occupationally-relevant exposures	Identify and assess workplaces where select engineered nanomaterials are manufactured or used and characterize potential worker exposures to those nanomaterials.

Participating federal agencies	Nonfederal partners	Date initiated	Type of collaboration	Title	Purpose
					Address the lack of information available on the identity and location of manufacture or use of engineered nanomaterials with high commercial potential and the lack of worker exposure data. Conduct 12 field investigations at sites where engineered nanomaterials are handled and manufactured. The focus of these investigations is to be on nanomaterials of interest to NIEHS and NIOSH. Give priority to materials on the Organisation for Economic Co-operation and Development's list of representative manufactured nanomaterials as well as those that are receiving increased attention due to requests to manufacture, use, or distribute in the United States.

Table 6. (Continued)

Participating federal agencies	Nonfederal partners	Date initiated	Type of collaboration	Title	Purpose
NIEHS, NIOSH	NA	March 2011	Interagency agreement	Durability of nanoscale cellulose fibers in artificial human lung fluids	Characterize the physicochemical properties and dissolution behaviors of nanoscale cellulose fiber materials in artificial lung fluids to determine their biodurability.
NIOSH	Institute of Occupational Medicined	March 2009	Collaborative research	Risk assessment of engineered nanoparticles	Conduct short-term inhalation experiments in mice to investigate short-term kinetics and health effects (e.g., pulmonary, cardiovascular, and central nervous system responses) of one or two selected engineered nanoparticles. Develop risk assessment methods, including biomathematical models of exposure-dose response, sensitivity and uncertainty analyses, and extrapolation from *in vivo*studies in rodents to predict internal dose and risk in humans.

Participating federal agencies	Nonfederal partners	Date initiated	Type of collaboration	Title	Purpose
					Provide exposure datain workplaces where nanoparticles are produced or used and develop risk management strategies.
NIOSH	University of Massachusetts, Lowell	June 2009	Contract	Professional services for developing a guidance document: *Safe Practices for Working with Engineered Nanomaterials in Research Laboratories*	Review the available scientific literature, conduct empirical measurements in laboratories, and apply best professional judgment to the development of a draft guidance document on safe practices for engineered nanomaterials in research laboratories.
NIOSH	Rice University's International Council on Nanotechnology (ICON)	August 2009	Contract	Development and deployment of a wiki to share good risk management practices for engineered nanomaterials	Develop a workspace for good risk management practice information, housed on the ICON GoodNanoGuide wiki. Using wiki technology for sharing risk management information, successful practices, and case studies, seek to provide NIOSH stakeholders and the public with an easy to use, familiar, flexible, and comprehensive platform to obtain critical occupational safety and health information.

Table 6. (Continued)

Participating federal agencies	Nonfederal partners	Date initiated	Type of collaboration	Title	Purpose
					Discussion forums can be established for users to collaborate on a wide range of topics, including novel facility design solutions, improvedcontrol technologies, or creative risk characterization techniques. New users can engage the experienced user community for nanomaterial management advice. Developing a wiki-based information sharing platform through ICON has the potential for reducing barriers to providing information to a government site.
NIOSH, Department of Defense (DOD)	NA	October 2009	Memorandum	Examination of potential occupational exposures of engineered nanomaterial at the DOD: a preliminary framework for DOD and NIOSH	Evaluate environment, safety, and occupational health issues related to the creation, handling, and use of engineered nanomaterials at facilities across all DOD agencies. This partnership can be effected across the DOD or on an individual basis

Participating federal agencies	Nonfederal partners	Date initiated	Type of collaboration	Title	Purpose
					(Department of Army, Department of Navy, etc.), dependent upon the degree of interest and need for such environmental, safety, and occupational health evaluations.
NIOSH	Battelle Memorial Institute Pacific Northwest Divisione	March 2010	Memorandum of understanding	Memorandum of understanding between Battelle Memorial Institute Pacific Northwest Division and NIOSH	Combine the capabilities of NIOSH and Battelle to advance the fundamental science supporting the safety assessment of nanotechnologies. Specifically, NIOSH and Battelle seek to engage in a joint research program studying the comparative toxicity of commercially important nanomaterials (e.g., single-walled carbon nanotubes and silica).
NIOSH	Center for High-Rate Nanomanufacturing at the University of Massachusetts, Lowell	August 2010	Contract	Engineering controls for nanomaterial handling	Use data from field research to evaluate the performance of available engineering controls and promote their use for nanomaterial manufacturing and handling. Develop guidance for safely handling nanomaterials.

Table 6. (Continued)

Participating federal agencies	Nonfederal partners	Date initiated	Type of collaboration	Title	Purpose
NIOSH	ICON	August 2010	Contract	Develop and deploy a wiki to share good risk management practices for engineered nanomaterials: follow-on deployment and evaluation	Perform follow-on deployment and evaluation tasks of work from previously completed contract. The objective of the platform, now known as the GoodNanoGuide, is to contribute to safety and health excellence by creating a generally accessible repository of critical safety and health research findings and best practice information toassist users in developing and administering effective safety and health programs to be used during the research, production, and use of engineered nanomaterials. GoodNanoGuide content is being contributed from a variety of sources, including private, government, and academic safety, health, and environmental practitioners and from the various national and international consortia

Participating federal agencies	Nonfederal partners	Date initiated	Type of collaboration	Title	Purpose
					that have been created to foster good practices. The long-range intent is to have a virtual warehouse (or library) where health, safety, and environmental professionals and the public can go to share information and collaborate on particular topics.
NIOSH, Occupational Safety and Health Administration	NA	August 2010	Interagency agreement	Guidance for nanomaterials workers	Investigate and characterize potential hazards associated with manufactured nanomaterials, with the goals of developing and recommending occupational safety and health measures to protect workers who manufacture and handle these materials. Build upon ongoing collaboration to investigate a control-focused guidance strategy for assisting small business employers in identifying and controlling workplace chemical hazards.
NIOSH	Lovelace Respiratory Research Institutef	September 2010	Contract	Deposition of carbon nanotubes in a human respiratory tract replica	Obtain measurement data on the deposition efficiency and size distribution of multi-walled carbon nanotubes

Table 6. (Continued)

Participating federal agencies	Nonfederal partners	Date initiated	Type of collaboration	Title	Purpose
					and single-walled carbon nanotubes in a human respiratory tract replica. These data are to be used by NIOSH to calibrate and evaluate a preliminary biomathematical model to predict the deposition of airborne carbon nanotubes in the human respiratory tract. A key element of this research is to develop an approach to relate the carbon nanotubes particle morphology to their aerodynamic and diffusion diameters.
NIOSH	University of Massachusetts, Lowell, the Center for High-rate Nanomanufacturing	September 2010	Memorandum of understanding	NIOSH-Center for High-rate Nanomanufacturing partnership to advance research and guidance for occupational safety and health in nanotechnology	Advance workplace health and safety standards and practices and strengthen the resulting nanotechnology workforce. The Center for High-rate Nanomanufacturing, as a part of this partnership, seeks to collaborate with NIOSH to provide a state-of-the-art platform for the discovery and dissemination of fundamental

Participating federal agencies	Nonfederal partners	Date initiated	Type of collaboration	Title	Purpose
					knowledge in the emerging interdisciplinaryfields of nanotechnology, including nanoscience, nanoengineering, and nanobiotechnology for the purpose of jointly developing effective occupational safety and health guidance.
NIOSH	College of Nanoscale Science and Engineering of the University at Albany—State University of New York	October 2010	Memorandum of understanding	NIOSH-College of Nanoscale Science and Engineering of the University at Albany partnership to advance research and guidance for occupational safety and health in nanotechnology	Establish a world-class intellectual and physical platform for occupational safety and health research, development, educational, and business initiatives leading to the discovery, demonstration, and deployment of new guidelines, recommendations, and findings relating to the potential human health impacts of exposure to engineered nanomaterials.
NIOSH	Applied Research Associates, Inc.g	March 2011	Contract	Development of a preliminary biomathematical model to predict deposition of carbon nanotubes and nanofibers in the respiratory tract of workers	Extend the carbon nanotubes deposition model to accommodate additional input data on both aerodynamic and diffusion diameters, and on particle number concentration as well as mass concentration.

Table 6. (Continued)

Participating federal agencies	Nonfederal partners	Date initiated	Type of collaboration	Title	Purpose
					An acceptable extended model is to provide predictions of the deposition of carbon nanotubes of different sizes and shapes (i.e., single-walled carbon nanotubes and multi-walled carbon nanotubes) in the human respiratorytract. The model is to be calibrated and evaluated—evaluation includes determining the particle properties and model parameters that provide the best predictions of multi-walled carbon nanotubes and single-walled carbon nanotubes deposition in the human respiratory tract replica. The carbon nanotubes deposition model is to be linked with an initial model to predict carbon nanotubes clearance and retention in the respiratory tract.

Participating federal agencies	Nonfederal partners	Date initiated	Type of collaboration	Title	Purpose
NIOSH	ICON	August 2011	Contract	Develop content and evaluate dissemination through the GoodNanoGuide wiki	Develop focused content that includes new work practices and case studies, moderating the wiki, providing user support, and conducting an evaluation of the effectiveness of the GoodNanoGuide.
NIOSH, NIST, DOD, Department of Energy	Over 20 private sector entities	Program first initiated in July 2005; multiple site visits occurred between February 2008 and October 2011	Letter of introduction	Site visits by NIOSH for the purpose of evaluating engineered nanomaterials in the workplace	Enter nanomaterial manufacturing facilities to characterize process and evaluate the potential for worker exposure to nanoparticles. NIOSH may publish information from the overall study in the primary literature, and may supplement its *Approaches to Safe Nanotechnology*document on its website to incorporate,in a de-identified manner, observations or findings specific to work practices, exposure controls, or risk management practices that come from the visit that maybe of benefit to the rest of the nanomaterial manufacturing community.

Table 6. (Continued)

Participating federal agencies	Nonfederal partners	Date initiated	Type of collaboration	Title	Purpose
NIST, NIOSH, NIH	Agreement with International Alliance for NanoEHS Harmonization;hother partners for the alliance also include over 10 private entities and international organizations; Collaborators include the European Commission's Institute for Reference Materials and Measurements	October 2008	Participation agreement	International Alliance for NanoEHS Harmonization	Create protocols for a limited number of *in vitro*and *in vivo*toxicology tests on a small number of representative nanoparticles to enable a "round robin" study with the goal to achieve inter-laboratory reproducibility of results and verify and/or validate predictability of *in vitro*assays for *in vivo*responses. This is to involve at least the use of nanomaterials and biologicals from a single source, and a single set of protocols. The implementation of a round robin laboratory set of tests based on these protocols. Ongoing and further development of protocols that take into consideration study design and methodology, nanomaterial physicochemical properties,

Participating federal agencies	Nonfederal partners	Date initiated	Type of collaboration	Title	Purpose
					and quantitative structure function relationships at the nanobio interface for *in vitro*and *in vivo*protocol development. A next phase is to involve round robin studies of common nanomaterials and biological systems, where independent laboratories are to choose a different set of agreed approaches to interrogate the biological end-points.

Source: GAO analysis based on formal collaborative EHS research project agreements provided by NNI member agencies' officials.

[a] This agreement funded a grant selected through the cooperative framework signed between EPA and the UK.

[b] The International Life Sciences Institute Research Foundation is a nonprofit organization based in Washington, D.C. that focuses on advancing the methods and application of science in risk assessment, human nutrition, and the prevention of obesity.

[c] SAIC-Fredrick is a wholly owned subsidiary of Science Applications International Corporation and operates exclusively under a single, long-term contract to the National Cancer Institute.

[d] The Institute of Occupational Medicine is an independent organization based in the UK that produces research, consulting, and other services related to occupational and environmental health, hygiene, and safety.

[e] Battelle Memorial Institute is an independent research and development organization based in Columbus, Ohio.

[f] The Lovelace Respiratory Research Institute is a private biomedical research organization dedicated to improving public health through research on the prevention, treatment, and cure of respiratory disease.

[g] Applied Research Associates, Inc. is a research and engineering company based in Albuquerque, New Mexico.

[h] International Alliance for NanoEHS Harmonization consists of internationally recognized experts drawn from academia and other scientific institutions from around the world. It is currently chaired by the Centre for BioNano Interactions at University College, Dublin.

End Notes

[1] Nanomaterials may also occur naturally, such as components of volcanic ash or ocean spray, or be created incidentally, including as byproducts of welding. Our review focuses on engineered nanomaterials. When the term nanomaterial is used in our report, it refers to engineered nanomaterials.

[2] Mihail C. Roco, Chad A. Mirkin, and Mark C. Hersam, eds., *Nanotechnology Research Directions for Societal Needs in 2020* (Netherlands: Springer, 2010).

[3] GAO, *Nanotechnology: Nanomaterials Are Widely Used in Commerce, but EPA Faces Challenges in Regulating Risk,* GAO-10-549 (Washington, D.C.: May 25, 2010).

[4] As of 2010, the NNI member agencies include the Departments of Defense, Education, Energy, Homeland Security, Justice, Labor, State, Treasury, and Transportation; the Department of Agriculture's National Institute of Food and Agriculture and Forest Service; the Department of Commerce's Bureau of Industry and Security, National Institute of Standards and Technology, and the U.S. Patent and Trademark Office; the Department of Health and Human Services' Food and Drug Administration, National Institute for Occupational Safety and Health, and the National Institutes of Health; the Department of the Interior's U.S. Geological Survey; the Consumer Product Safety Commission; Environmental Protection Agency; intelligence agencies; the International Trade Commission; National Aeronautics and Space Administration; Nuclear Regulatory Commission; National Science Foundation; Office of Management and Budget; and Office of Science and Technology Policy.

[5] The eight program component areas are (1) fundamental nanoscale phenomena and processes; (2) nanomaterials; (3) nanoscale devices and systems; (4) instrumentation research, metrology, and standards for nanotechnology; (5) nanomanufacturing; (6) major research facilities and instrumentation acquisition; (7) environment, health, and safety; and (8) education and societal dimensions.

[6] These three strategy documents are: (1) NSTC, Committee on Technology, Subcommittee on Nanoscale Science, Engineering, and Technology, *National Nanotechnology Initiative Strategic Plan* (February 2011); (2) NSTC, Committee on Technology, Subcommittee on Nanoscale Science, Engineering, and Technology, *National Nanotechnology Initiative Environmental, Health, and Safety Research Strategy* (October 2011); and (3) NSTC, Committee on Technology, Subcommittee on Nanoscale Science, Engineering, and Technology, *The National Nanotechnology Initiative Supplement to the President's 2012 Budget* (February 2011).

[7] These documents also describe mechanisms for outreach and collaboration with stakeholders who are involved in nanotechnology EHS research, including those working in state and local governments, academia, companies, and nongovernmental organizations.

[8] GAO, *National Capital Region: 2010 Strategic Plan is Generally Consistent with Characteristics of Effective Strategies,* GAO-12-276T (Washington, D.C.: Dec. 7, 2011); *Influenza Pandemic: Further Efforts Are Needed to Ensure Clearer Federal Leadership Roles and an Effective National Strategy*, GAO-07-781 (Washington, D.C.: Aug. 14, 2007); *Financial Literacy and Education Commission: Further Progress Needed to Ensure an Effective National Strategy*, GAO-07-100 (Washington, D.C.: Dec. 4. 2006); *Combating Terrorism: Evaluation of Selected Characteristics in National Strategies Related to Terrorism*, GAO-04-408T (Washington, D.C.: Feb. 3, 2004).

[9] See NSTC, Committee on Technology, Subcommittee on Nanoscale Science, Engineering, and Technology, *The National Nanotechnology Initiative Supplement to the President's 2012 Budget* (February 2011); *The National Nanotechnology Initiative Supplement to the*

President's 2011 Budget (February 2010); *The National Nanotechnology Initiative Supplement to the President's 2010 Budget* (May 2009); *The National Nanotechnology Initiative Supplement to the President's 2009 Budget* (September 2008); and *The National Nanotechnology Initiative Supplement to the President's 2008 Budget* (July 2007).

[10] The Biomedical Research and Development Price Index is prepared by the Department of Commerce, Bureau of Economic Analysis for use of and publication by NIH. Information about the index is available at http://officeofbudget.od.nih.gov/pdfs/FY13/ BRDPI_ Proj_Jan_2012_Final.pdf.

[11] GAO, *Nanotechnology: Better Guidance Is Needed to Ensure Accurate Reporting of Federal Research Focused on Environmental, Health, and Safety Risks*, GAO-08-402 (Washington, D.C.: Mar. 31, 2008).

[12] The five categories are carbon-based nanomaterials, metal-based nanomaterials, semiconductor-based nanomaterials, organic nanomaterials, and other nanomaterials.

[13] In this report, we did not attempt to summarize the current scientific literature on the potential hazards of or exposure to nanomaterials. We most recently reviewed what is known about the potential human health and environmental risks of nanomaterials in 2010, see GAO-10-549.

[14] NRC, *A Research Strategy for Environmental, Health, and Safety Aspects of Engineered Nanomaterials* (2012).

[15] GAO-10-549.

[16] The Organisation for Economic Co-operation and Development is an organization of 34 national governments, operating by consensus, that fosters dialogue among members to discuss, develop, and refine economic and social policies and provides an arena for setting rules when multilateral agreements are necessary.

[17] The nanomaterials the Organisation for Economic Co-operation and Development selected are fullerenes, single-walled carbon nanotubes, multi-walled carbon nanotubes, silver nanoparticles, iron nanoparticles, titanium dioxide, aluminum oxide, cerium oxide, zinc oxide, silicon dioxide, dendrimers, nanoclays, and gold nanoparticles.

[18] GAO, *Food Safety: FDA Should Strengthen Its Oversight of Food Ingredients Determined to Be Generally Recognized as Safe (GRAS)*, GAO-10-246 (Washington, D.C.: Feb. 3, 2010). Subsequent to this report, FDA issued draft guidance on April 20, 2012, on assessing the effects of significant manufacturing changes, including any involving nanotechnology, on the identity, safety, or regulatory status of food substances.

[19] GAO-10-549.

[20] Pub. L. No. 108-153 (2003).

[21] NSTC, Committee on Technology, Subcommittee on Nanoscale Science, Engineering, and Technology, *National Nanotechnology Initiative Strategic Plan* (February 2011).

[22] NSTC, Committee on Technology, Subcommittee on Nanoscale Science, Engineering, and Technology, *National Nanotechnology Initiative Strategy for Nanotechnology-Related Environmental, Health, and Safety Research* (February 2008) and *National Nanotechnology Initiative Environmental, Health, and Safety Research Strategy* (October 2011).

[23] NRC, *A Research Strategy for Environmental, Health, and Safety Aspects of Engineered Nanomaterials* (2012). We did not fully evaluate this strategy in the context of federal EHS research efforts because it was issued in draft form in January 2012.

[24] The working group invited additional experts to speak at public meetings, including individuals working in academia and the corporate sector.

[25] This includes funding by the seven agencies we selected (CPSC, EPA, FDA, NIH, NIOSH, NIST, and NSF) as well as by the other NNI members funding nanotechnology EHS research (the Departments of Defense, Energy, and Agriculture).

[26] Adjusted for inflation, the amount of nanotechnology EHS research funding NSTC reported for fiscal year 2006 was approximately $43 million, in 2010 dollars.

[27] This includes funding by the seven agencies we selected as well as by the other NNI members funding nanotechnology EHS research.

[28] GAO-08-402.

[29] OMB Circular No. A-11, Preparation, Submission, and Execution of the Budget, pt. 2, § 84 (July 2010).

[30] GAO-08-402.

[31] FDA, *Strategic Priorities 2011-2015.*

[32] OMB's data calls for 2007 to 2009 actual agency investments did not specify what type of budget data should be reported; the data call for 2010 data asked agencies to report budget authority for their actual agency investments.

[33] Budget authority is authority provided by federal law to enter into financial obligations that will result in immediate or future outlays involving federal government funds. Budget authority includes (1) appropriations, (2) borrowing authority, (3) contract authority, and (4) authority to obligate and expend offsetting receipts and collections. Appropriations represent budget authority to incur obligations and make payments from the Treasury for specified purposes. Appropriations do not represent cash actually set aside in the Treasury for purposes specified in the appropriation act; they represent amounts that agencies may obligate during the period of time specified in the respective appropriation acts. See GAO, *A Glossary of Terms Used in the Federal Budget Process*, GAO-05-734SP (Washington, D.C.: Sept. 1, 2005).

[34] Obligations are definite commitments that create a legal liability of the government for the payment of goods and services ordered or received, or a legal duty on the part of the United States that could mature into a legal liability by virtue of actions on the part of the other party beyond the control of the United States. Payment may be made immediately or in the future. An agency incurs an obligation, for example, when it places an order, signs a contract, awards a grant, purchases a service, or takes other actions that require the government to make payments to the public or from one government account to another. See GAO-05-734SP.

[35] As directed in the President's Memorandum on Transparency and Open Government issued on January 21, 2009, OMB issued an Open Government Directive instructing executive branch agencies to, among other things, publish government information online.

[36] However, the NNI Supplement to the President's 2011 Budget did include data on 2009 actual agency investments from Recovery Act appropriations.

[37] NIH officials told us that, for fiscal year 2010, NIH obligated funds appropriated by the Recovery Act as follows: PCA 1, $19.8 million; PCA 2, $15.6 million; PCA 3, $28.3 million; PCA 4, $3.9 million; PCA 7, $8 million; and PCA 8, $0.5 million.

[38] NIST officials told us that, for fiscal year 2010, NIST obligated funds appropriated by the Recovery Act as follows: PCA 5, $3.1 million; and PCA 6, $34.3 million.

[39] NSF officials told us that, for fiscal year 2010, NSF obligated $17.7 million for PCA 6 research with funds appropriated by the Recovery Act.

[40] We did not determine the amounts of 2010 investments from Recovery Act appropriations, if any, by the NNI member agencies funding nanotechnology research that were not included

in this review (the National Aeronautics and Space Administration and the Departments of Defense, Energy, Agriculture, Justice, Homeland Security, and Transportation).

[41] For example, according to NIOSH officials, the agency does not report to OMB funds that provide support for major program areas that require shared services and facilities, because of the difficulty of placing portions of a central support project into a program component area. Also, the automated method NIH uses to compile its list of nanotechnology EHS research projects may not identify all appropriate projects because it relies on a computerized search of key words, which will not define an absolute dividing line between projects that should be included in the EHS program component area and those that should not. According to NIH officials, this process provides a consistent, reliable, and repeatable method for efficiently scanning NIH's projects, which number more than 50,000 annually. We did not review information on NIH projects that were not reported to us by NIH as nanotechnology EHS research. However, in reviewing information on the 58 fiscal year 2010 nanotechnology EHS research projects NIH reported to us, we identified one example of funding for a subproject that was missed by this method.

[42] The numbers of projects do not total 193 because (1) some projects targeted nanomaterials other than carbon nanotubes, nanosilver, and nanoscale titanium dioxide; (2) projects often targeted multiple nanomaterials; and (3) nineteen of the 193 projects were not targeted to particular nanomaterials. For example, some projects not targeted to particular nanomaterials funded large research centers on the environmental implications of nanotechnology; others purchased data on the market for nanomaterials to guide the agency's exposure assessment studies by providing information on the potential consumer exposures.

[43] NIOSH, *Approaches to Safe Nanotechnology: Managing the Health and Safety Concerns Associated with Engineered Nanomaterials*, Department of Health and Human Services (NIOSH) Publication No. 2009–125 (March 2009).

[44] EPA, *State of the Science Literature Review: Everything Nanosilver and More*, EPA/600/R-10/084 (August 2010).

[45] NIOSH, *Current Intelligence Bulletin 63: Occupational Exposure to Titanium Dioxide*, Department of Health and Human Services (NIOSH) Publication No. 2011–160 (April 2011).

[46] N. Sadrieh, A.M. Wokovich, N.V. Gopee, J. Zheng, D. Haines, D. Parmiter, P.H. Siitonen, C.R. Cozart, A.K. Patri, S.E. McNeil, P.C. Howard, W.H. Doub, L.F. Buhse, "Lack of Significant Dermal Penetration of Titanium Dioxide from Sunscreen Formulations Containing Nano- and Sub-Micron-Size TiO2 Particles," *Toxicological Sciences,* vol. 115, no. 1 (2010):156-66.

[47] GAO-10-549.

[48] NRC, *A Research Strategy for Environmental, Health, and Safety Aspects of Engineered Nanomaterials* (2012).

[49] The Wilson Center is a nonpartisan research institution established by an act of Congress in 1968 and supported by public and private funds. The Project on Emerging Nanotechnologies was established in April 2005 as a partnership between the Wilson Center and the Pew Charitable Trusts. The project's mission is to, among other things, collaborate with researchers, government, industry, and others to identify gaps in knowledge and regulatory processes, develop strategies for closing them, and provide independent, objective knowledge and analysis related to the development and commercialization of nanotechnologies.

[50] According to the Wilson Center, the inventory is based on information that can be readily found on the Internet about products that (1) can be readily purchased by consumers and (2) are identified as nanotechnology-based by the manufacturer or another source. The center tried to avoid including products that clearly do not use nanotechnology, but they did not verify manufacturers' claims or conduct any independent testing of products.

[51] GAO-10-549.

[52] The IANH is a group of internationally recognized experts who have agreed to develop specific tools and testing protocols and perform a set of round robin experiments for reproducible testing of nanomaterial biological interactions and toxicology. The International Life Sciences Institute Research Foundation is a nonprofit organization based in Washington, D.C., that focuses on advancing the methods and application of science in risk assessment, human nutrition, and the prevention of obesity.

[53] The NSTC finalized the 2011 NNI EHS research strategy while the questionnaire was being completed by respondents. Therefore, some respondents may have rated the draft 2011 NNI EHS research strategy while others may have rated the final version.

[54] NSTC, Committee on Technology, Subcommittee on Nanoscale Science, Engineering, and Technology, *National Nanotechnology Initiative Environmental, Health, and Safety Research Strategy* (October 2011); *National Nanotechnology Initiative Strategic Plan* (February 2011); and *The National Nanotechnology Initiative Supplement to the President's 2012 Budget* (February 2011).

[55] The "partially addresses" category includes a range that varies from explicitly citing most of the elements to citing as few as one of the elements of a characteristic. Yet because the three NNI documents have different purposes and audiences an individual document need not address each desirable national strategy characteristic. However, areas not addressed in one document should be more fully developed in another.

[56] GAO-04-408T.

[57] CPSC, *2010 Performance and Accountability Report* (November 2010); Department of Health and Human Services, Food and Drug Administration, *FY2012 Online Performance Appendix*; Department of Health and Human Services, National Institutes of Health, *FY2012 Online Performance Appendix*.

[58] EPA, *Nanomaterial Research Strategy*, EPA 620/K-09/011 (June 2009); NIOSH, *Strategic Plan for NIOSH Nanotechnology Research and Guidance*, Department of Health and Human Services (NIOSH) Publication No. 2010-105 (November 2009); NIST, *Nanomaterial Environmental, Health and Safety (Nano-EHS) Program Plan, FY10-FY14*.

[59] GAO, *Results-Oriented Government: Practices That Can Help Enhance and Sustain Collaboration Among Federal Agencies*, GAO-06-15 (Washington, D.C.: Oct. 21, 2005).

[60] The 2011 NNI EHS research strategy states that, because individual agency priorities may differ in scope and focus from the NNI research needs, the NSTC relies on coordination through NEHI to ensure integration of agency plans.

[61] FDA, Advancing Regulatory Science at FDA (August 2011), http://www.fda.gov/ScienceResearch/SpecialTopics/RegulatoryScience/ucm267719.htm, and FDA, Nanotechnology Regulatory Science Research Plan, http://www.fda.gov/ScienceResearch/SpecialTopics/Nanotechnology/ucm273325.htm.

[62] NSF, NSF FY2012 Budget Request to Congress, http://www.nsf.gov/about/budget/fy2012/index.jsp.

[63] GAO-08-402.

End Notes for Appendix I

[1] See NSTC, Committee on Technology, Subcommittee on Nanoscale Science, Engineering, and Technology, The National Nanotechnology Initiative Supplement to the President's 2012 Budget (February 2011); The National Nanotechnology Initiative Supplement to the President's 2011 Budget (February 2010); The National Nanotechnology Initiative Supplement to the President's 2010 Budget (May 2009); The National Nanotechnology Initiative Supplement to the President's 2009 Budget (September 2008); and The National Nanotechnology Initiative Supplement to the President's 2008 Budget (July 2007).

[2] GAO, Nanotechnology: Better Guidance Is Needed to Ensure Accurate Reporting of Federal Research Focused on Environmental, Health, and Safety Risks, GAO-08-402 (Washington, D.C.: Mar. 31, 2008).

[3] NSTC, Committee on Technology, Subcommittee on Nanoscale Science, Engineering, and Technology, National Nanotechnology Initiative Strategy for Nanotechnology-Related Environmental, Health, and Safety Research (February 2008).

[4] A nonprobability sample is a sample in which some items in the population have no chance, or an unknown chance, of being selected. Results from nonprobability samples cannot be used to make inferences about a population. However, information from the sample provided illustrative examples.

[5] In 2009 and 2010, the NNI held four public workshops: (1) Human and Environmental Exposure Assessment (February 2009); (2) Nanomaterials and the Environment & Instrumentation, Metrology, and Analytical Methods (October 2009); (3) Nanomaterials and Human Health & Instrumentation, Metrology, and Analytical Methods (November 2009); and (4) Risk Management Methods & Ethical, Legal, and Societal Implications of Nanotechnology (March 2010).

[6] For example, administrative assistants or staff managers were excluded because they are unlikely to have subject matter expertise. In another case, an individual in a consultancy that does not focus on nanotechnology was excluded, based on advice from the head of a nanotechnology trade association.

[7] Not all individuals responded to every question.

[8] These three strategic documents are: (1) NSTC, Committee on Technology, Subcommittee on Nanoscale Science, Engineering, and Technology, National Nanotechnology Initiative Strategic Plan (February 2011); (2) NSTC, Committee on Technology, Subcommittee on Nanoscale Science, Engineering, and Technology, National Nanotechnology Initiative Environmental, Health, and Safety Research Strategy (October 2011); and (3) NSTC, Committee on Technology, Subcommittee on Nanoscale Science, Engineering, and Technology, The National Nanotechnology Initiative Supplement to the President's 2012 Budget (February 2011).

[9] GAO, National Capital Region: 2010 Strategic Plan is Generally Consistent with Characteristics of Effective Strategies, GAO-12-276T (Washington, D.C.: Dec. 7, 2011); Influenza Pandemic: Further Efforts Are Needed to Ensure Clearer Federal Leadership Roles and an Effective National Strategy, GAO-07-781 (Washington, D.C.: Aug. 14, 2007); Financial Literacy and Education Commission: Further Progress Needed to Ensure an Effective National Strategy, GAO-07-100 (Washington, D.C.: Dec. 4, 2006); Combating Terrorism: Evaluation of Selected Characteristics in National Strategies Related to Terrorism, GAO-04-408T (Washington, D.C.: Feb. 3, 2004).

[10] GAO-07-781.

INDEX

D

H

I

J

K

L

M

N

O

P

Q

R

S

T

U

V

W

Y

Z